NOTES ON
METAL COATING
TECHNOLOGY

by

Henry Leidheiser, Jr.

Lehigh University
Bethlehem, Pennsylvania

Second Revised Edition

William Kornthauer

Wexford Press
2009

TABLE OF CONTENTS

CONTENTS (Cont'd.)

Section 1

Useful Literature Sources

Books

"The Properties of Electrodeposited Metals and Alloys. A Handbook," W. H. Safranek, American Elsevier Publ. Co., 517 pp. (1974).

"Metal Finishing Guidebook and Directory," published annually by Metals and Plastics Publ., Inc., Hackensack, N.J.

"Metallic Coatings for Corrosion Control," V. E. Carter, Newnes-Butterworths, 183 pp. (1977).

"Design and Corrosion Control," V. R. Pludek, Halsted Press, 383 pp. (1977).

"Corrosion Engineering, 2nd Edition, M. G. Fontana and N. D. Greene, McGraw-Hill Book Co., 465 pp (1978).

Journals

Plating and Surface Finishing, published monthly by Am. Electroplaters Society

Metal Finishing, published monthly by Metals and Plastics Publ. Co.

Materials Performance, published monthly by National Assoc. of Corrosion Engineers

Corrosion, published monthly by National Assoc. of Corrosion Engineers

Surface Technology, published monthly by Elsevier Publ. Co.

Section 2

Metallic Coatings Commonly Used

Aluminum

Corrosion resistance to atmosphere is good with a rate of the order of 2-5 μm per year. Rate decreases with time of exposure.

Thin oxide film provides the corrosion resistance for the metal. When this oxide film is removed, aluminum is very reactive.

In the atmosphere, the attack of aluminum is generally in the form of pits which occur at weak points in the oxide film.

When immersed in an electrolyte, the rate of corrosion of aluminum is a function of the dissolved oxygen concentration, the chloride content and the presence of heavy metal ions such as copper. High oxygen concentrations in solution generally lead to a lower corrosion rate while high chloride contents and the presence of heavy metals lead to an increased rate of corrosion.

Things to be concerned with in the case of aluminum coatings

Crevice corrosion
Galvanic corrosion in contact with iron or steel
Pitting
Run-off water passing over copper
Aluminum alloys containing copper, zinc and magnesium are
    susceptible to stress corrosion cracking

TYPICAL APPLICATIONS OF ALUMINIUM COATINGS

Substrates	Applications	Coating methods
Steel	Structure exposed to the atmosphere, immersed in water or buried. Components to resist high-temperature oxidation or hot flue gases	Hot-dipping, or metal spraying, or cladding
	Decorative finishes	Vacuum deposition
Aluminium alloys	Structures or components exposed to aggressive atmospheres, immersed in water or buried (applicable particularly to the protection of aluminium alloys from stress-corrosion)	Metal spraying or cladding
Plastics	Decorative finishes (particularly for reflective finishes)	Vacuum deposition

Cadmium

Corrosion resistance to atmosphere is fair. In severe industrial atmosphere, a 25 μm thick coating will protect steel substrate for a year or more. The lifetime is greater in a marine environment. Cadmium sulfates formed in industrial atmosphere are soluble and are washed away by rain; cadmium carbonates and basic cadmium chlorides are less soluble and are retained on surface and act as a partial barrier to further corrosion.

Cadmium coatings are more tarnish resistant than zinc.

Advantages in Use of Cadmium

    Bright appearance
    Ease of soldering
    Slight resistance to alkali

Disadvantages in Use of Cadmium

    Not as effective as zinc as a sacrificial anode
    Expensive
    Salts and vapor are toxic
    Sensitivity to inducing hydrogen embrittlement when plated on high
        strength steel
    Low strength

TYPICAL APPLICATIONS OF CADMIUM COATINGS

Substrates	Applications	Coating methods
Steel	Structures and fasteners exposed to humid atmospheres or to organic vapours. Surfaces requiring good solderability. Low-torque threaded fasteners. Components in bimetallic contact with aluminium	Electrodeposition or vacuum deposition

Chromium

Two major uses are in decorative coatings and in wear-resistant coatings.

Thin decorative chromium deposits are plated over nickel or over a combination of copper and nickel. The chromium deposits are always porous and exhibit cracks because of internal stresses formed during electrodeposition. Thickness is of the order of 0.3 μm.

Hard chromium deposits range in thickness from 10 μm to 400 μm. The deposits contain cracks but these cracks serve to retain lubricant during service.

Chromium deposits are not useful in very aggressive environments since the porosity of the coating allows attack of the substrate.

TYPICAL APPLICATIONS OF CHROMIUM COATINGS

Substrates	Applications	Coating methods
Steel	Decorative overlay to protective nickel coatings on components exposed to the atmosphere. Wear-resistant coatings on engineering components, e.g. rollers, hydraulic rams, printing cylinders	Electrodeposition
	Hard, wear-resistant coatings on engineering components. Coatings resistant to high-temperature oxidation	Diffusion coating
Aluminium, copper and its alloys, zinc alloys	Decorative overlay applied either directly or over protective nickel undercoats on components exposed to the atmosphere	Electrodeposition

Copper

Copper and copper alloys have high resistance to atmosphere, with corrosion rates of 0.2-0.6 μm per year in a rural environment and 0.9-2.2 μm in an industrial environment. Corrosion leads to the green patina so often seen on old copper roofs.

Copper can be used in seawater and imbedded in soils.

Things to be Concerned about in Use of Copper and its Alloys

Ammonia environments are to be avoided. Such environments lead to corrosion of copper and to stress corrosion cracking of copper alloys.

Copper alloys are subject to erosion when used in condenser systems and pitting may occur.

Copper is cathodic to steel; breaks in the coating on steel may lead to rapid corrosion of the steel.

TYPICAL APPLICATIONS OF COPPER COATINGS

Substrates	Applications	Coating methods
Steel	Decorative and protective coatings resistant to atmospheres or water immersion. Surfaces requiring good solderability. Surfaces requiring good electrical conductivity	Electrodeposition, or electroless plating, or cladding
Steel or zinc alloys, aluminium	Undercoat for protective nickel/chromium coatings. Engineering coatings for printing, engraving or electronic applications	Electrodeposition or electroless plating
Plastics	Preliminary coatings for protective plating of plastics	Electroless plating and electrodeposition

Gold

Resistant to corrosion in all but the most aggressive environments as aqua regia, cyanide solutions, and liquid mercury.

Generally used as a decorative coating or a protective coating for electronic components and connectors.

Gold has very good electrical conductivity and because of its resistance to corrosion and tarnishing it retains a low electrical contact resistance for long periods of time.

Gold is soft and is easily scratched or mechanically deformed. Cobalt salts are added to plating bath to obtain a hard deposit.

TYPICAL APPLICATIONS OF GOLD COATINGS

Substrates	Applications	Coating methods
Copper and its alloys	Decorative and protective coatings for jewellery. Protective coatings for aerospace hardware. Protective coatings with good electrical conductivity for electronic applications	Electrodeposition, or electroless plating, or cladding, or vacuum deposition
Plastics	Electrically conducting coatings	Electroless plating or vacuum deposition

Indium

Indium melts at 155°C and is so soft that it has a greasy feel. It is most widely used for plating and diffusion into thin lead sleeve bearing surfaces. It can be diffused into copper alloys to produce a wear-resistant, buffable surface with increased tarnish resistance. It can also be diffused below its melting point into lead, tin, cadmium, zinc, silver and gold.

Indium is also applied as a preplate for rhodium in order to reduce the rhodium thickness necessary for tarnish resistance.

Lead

Lead exhibits good corrosion resistance in industrial atmospheres, in soils, and in waters. Many of the compounds of lead are insoluble and these insoluble products protect the lead from further corrosion. Lead is resistant to attack by concentrated sulfuric acid.

Lead has extreme softness and ductility. These features can be both an advantage and a disadvantage in protective applications.

Lead is a good material for sound dampening.

Lead compounds are toxic.

TYPICAL APPLICATIONS OF LEAD COATINGS

Substrates	Applications	Coating methods
Steel or copper	Acid-resistant coatings for chemical plant. Structures resistant to atmospheres, waters or buried conditions. Surfaces requiring good solderability. Sound-damping applications	Hot-dipping, or cladding, or electrode position

Nickel

High degree of corrosion resistance to the atmosphere and to waters. Corrosion rate in the atmosphere is in the range of 0.02 - 0.2 µm per year.

Nickel electrodeposits serve primarily as barrier layers; they offer no sacrifical protection to steel or zinc alloy substrates. Corrosion occurs at breaks in the nickel and the deposit is blistered away as the corrosion occurs.

Nickel is useful for elevated temperature exposures because of the low rate of thickening of the oxide.

Nickel electrodeposition solutions can be designed so as to deposit bright coatings, hard and brittle coatings, ductile coatings, leveling of scratches and projections in the substrate, and with variable rates of corrosion.

Substrates	Applications	Coating methods
Steel or zinc alloys, copper and its alloys, aluminium	Corrosion-resistant coatings for components or structures exposed to atmospheres or immersed in waters or sea water (used either alone or in combination with copper undercoats and/or chromium overlays). Protective coatings for chemical plant. Hard, wear-resistant coatings for engineering applications	Electrodeposition, or metal spraying, or cladding, or electroless plating
Plastics	Preliminary coatings for protective plating of plastics	Electroless plating and electrodeposition

High nickel alloys, such as Inconel, are resistant to SCC.

Precious Metals

The precious metals such as palladium, platinum, rhodium and iridium can be plated as thin films. Their major applications are in jewelry and to a limited extent in semiconductor device protection. The deposits often have inferior mechanical properties. Rhodium is used to coat silver when used as a mirror on glass. Platinum is used in the plating of titanium anodes which are used in acidic solutions or as anodes in galvanic protection systems.

Silver

Silver has good electrical conductivity and good color for decorative purposes, but it readily tarnishes in atmospheres containing sulfur compounds.

Appearance can be maintained by overcoating with a thin layer of rhodium.

TYPICAL APPLICATIONS OF SILVER COATINGS

Substrates	Applications	Coating methods
Copper and its alloys, nickel and its alloys	Decorative coatings for jewellery, cutlery and household articles. Protective coatings for chemical plant. Components for electronic applications	Electroless plating, or electrodeposition, or vacuum deposition
Plastics	Decorative coatings. Coatings for good electrical conductivity	Electroless plating or vacuum deposition

Tin

Tin is very resistant to atmospheric attack. It is virtually unattacked by immersion in potable waters and only very slowly attacked by sea water.

Tin is very soft and ductile. It melts at $232^{\circ}C$ and thus is not useful for high temperature applications. Tin has good solderability and good electrical conductivity.

It may transform to a low temperature form upon continued exposure to very low temperatures. This transformation, called "tin disease," is rare and is generally not a consideration in normal applications of tin coatings.

TYPICAL APPLICATIONS OF TIN COATINGS

Substrates	Applications	Coating methods
Steel, copper and its alloys	Protective coatings for resistance to atmospheres, immersion in water or organic acids. Food canning. Surfaces requiring good solderability and electrical conductivity	Hot-dipping or electrodeposition

Zinc

The corrosion rate of zinc is about 15 μm per year in an industrial atmosphere. It falls to about 1/5 of this rate in rural or marine environments. The corrosion rate is constant and the lifetime of a coating is directly proportional to its thickness.

The major advantage of zinc is the sacrificial protection it provides to steel. At breaks in the coating, the steel becomes the cathode and the zinc corrodes in preference to the steel.

Attack of zinc is greatly accelerated by contact with metals such as copper or noble metals such as platinum.

Zinc is attacked both by acidic and alkaline environments and finds its greatest usefulness in environments that are near neutral.

TYPICAL APPLICATIONS OF ZINC COATINGS

Substrates	Applications	Coating methods
Steel, aluminium alloys	Protective coatings for structures and fasteners or components exposed to atmospheres, immersed in water or buried	Hot-dipping, or electrodeposition, or metal spraying, or vacuum deposition

Alloy Coatings

Aluminum-Zinc-Silicon. This alloy is a recent commercial development and is marketed as a substitute for galvanized steel. The tradename is "Galvalume." It has higher resistance to atmospheric attack than pure zinc. It is produced by a hot dipping process.

Copper-Tin. An alloy containing 88% copper and 12% tin is electroplated as a decorative coating.

Copper-Zinc. Brass is an alloy of copper and zinc and is the most widely used alloy electrodeposit. The largest amount of brass plating is for decorative use but there are important engineering uses such as the plating of steel wire cord for steel-belted radial tires. Brass serves as a good bonding agent to rubber.

Lead-Tin. An alloy containing 8-12% tin is known as terne is formed as a coating by hot dipping. The corrosion resistance is good and terne coated steel is used for the interior of gasoline tanks. The coating has good solderability, drawability, paintability, and weldability.

Tin-Cobalt. Electrodeposited tin-cobalt alloys have been used as a substitute for decorative chromium.

Tin-Lead Alloys. Ranging from 60% tin to 95% tin are normally used as a coating for improved corrosion protection and as a base for soldering. They are electrodeposited.

Tin-Nickel. This alloy containing about 50 atomic percent tin and 50 atomic percent nickel is very corrosion resistant and has found limited application as a protective coating for electronic devices. They are generally electrodeposited.

Aluminum-Zinc. Metal spraying.

Section 3

Surface Preparation Prior to the Application of the Coating

Note: This section has been taken verbatim from the book, "Metallic Coatings for Corrosion Control" by V. E. Carter.

Before any process of metallic coating is applied to a metal substrate it is essential that the latter's surface shall be in a suitable condition to receive the coating. In order to achieve this one or more pretreatment processes must be employed. Broadly, pretreatments fulfil one or more of three purposes:

(a) removal of surface contaminants
(b) removal of superficial corrosion
(c) control of the physical nature of the metal surface

The choice of any or all of these types of process and of the order in which they are applied depend upon the condition of the substrate material as received, on the type of coating process that is subsequently to be used and on the end-use of the coated article.

Because of these differing pretreatment procedures and the factors affecting the choice of specific ones for a given purpose the "pretreatment line" preceding a coating process can range from a single simple operation to a complex multi-process sequence. It is not the purpose of this section to set out in detail the exact requirements of pretreatment lines for particular products and coating processes, but rather to give the reader an outline of the various methods used, the reasons for their use and the ways in which they are applied to meet the requirements of the three classifications given above.

Removing surface contaminants

Surface contaminants are almost always present on materials as a result of production processes that have been carried out prior to receipt, or as a result of deliberate application in order to provide temporary protection or identification. They are, usually, primarily of an organic nature--oils, greases, waxes, paints, lacquers, etc.--but may also be combined with inorganic materials such as metallic debris from the bulk metal produced during mechanical working operations (e.g. swarf or metal soaps) and particulate dirt derived from airborne pollutants.

The presence of surface contaminants always seriously hinders the successful application of a coating process for the following reasons.

(a) They can scar the metal surface during any polishing treatments that may be required, or may even be driven into the surface of the metal so that they cannot easily be removed.

(b)   They can provide a physical barrier that will prevent access of a processing solution to the metal surface so that the requisite reaction cannot occur.

(c)   They can react with a processing solution, altering its chemical composition and hence its reactions with the metal to be coated.

(d)   In the presence of an electrolyte (such as a processing solution) they can react with the substrate metal or with the coating metal, causing corrosion of the surface or producing insoluble products that will further contaminate the surface.

(e)   They can be incorporated in a coating system, producing a region where coating adhesion may be defective or interfering with the homogeneity or growth of the coating itself so that a physical defect may develop. Any such defects produced during the early stages of the processing sequence may provide a region where subsequent processing solutions can become entrapped; these pockets of entrapped solution can themselves produce corrosion at a later time.

It follows from the above possibilities that surface contaminants should always be removed before subsequent processing is attempted. The principal way in which contaminants of this nature are removed from metals is by the use of cleaner-degreasers.

In its simplest form a cleaner-degreaser may be merely a tank of organic solvent (such as carbon tetrachloride, benzene, toluene, acetone, etc.) maintained at room temperature, into which the work may be dipped or swabbed. Oils, greases and lacquers are softened by the solvent and taken into solution, and entrapped insoluble dirt and metal particles are loosened so that they can fall away to the bottom of the vat. However, simple immersion or swabbing in the cold is an inefficient means of cleaning all but limited quantities of small articles. Problems are associated with the extraction of toxic vapours from the solvent; also the vat quickly becomes contaminated with dirt and greases removed from the work, which form an emulsion that is retained as a film on the metal surface after removal and drying.

Heat can be applied to the bath to accelerate cleaning action, and some amelioration of bath contamination can be achieved by working a closed-circuit flow of solvent incorporating settling and/or filtering, but even these methods remain of limited efficiency.

The most commonly used method of solvent degreasing that operates at high efficiency is the hot liquid/vapour degreaser plant. The principle of this type of equipment is shown in Figure 3.1. The solvent is contained in a tank in which it can be heated to boiling point. There is an annular compartment in the upper part of the side walls of the tank within which is a cooling coil where the hot solvent vapour is condensed back into liquid. This condensate is collected in the bottom of the annular compartment, from which it flows back by gravity feed into the main liquid compartment below to recommence the vaporizing-condensing cycle. The tank is covered and vented through a vertical flue, which extracts any uncondensed fumes that may escape from the upper portion of the tank.

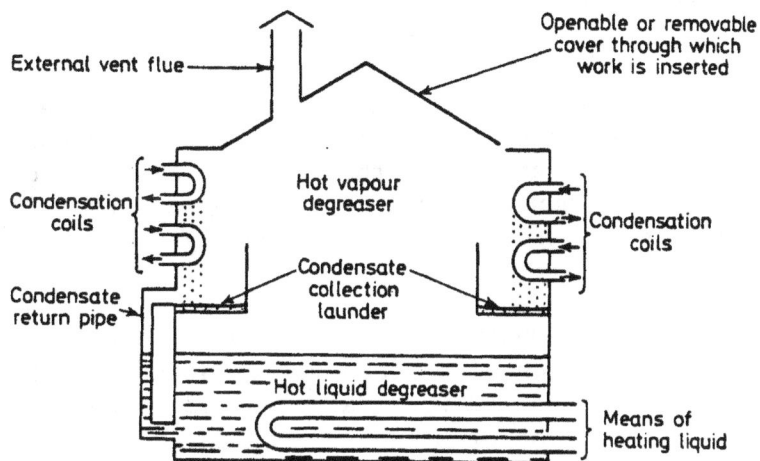

Figure 3.1. *Diagrammatic sketch of hot liquid/vapour degreaser*

If cold, soiled work is introduced into the upper part of the chamber where the solvent vapour is present at a temperature of, say, 87°C (the boiling point of trichloroethylene), condensation of the solvent takes place on its cool surface. This continues until the temperature of the work attains that of the solvent vapour. During this period the continuously renewed flow of condensate on the surface of the work flushes away soils and greases, which fall to the bottom of the tank. If the work is heavily soiled with resistant contaminants, treatment in this way by solvent condensate may be insufficient to effect complete cleaning. In such cases the work may be totally immersed in the boiling solvent tank in which the heavy soils will be removed with a high degree of efficiency. After total immersion a light grease film may be retained on draining, cooling and drying, but this can readily be removed by a subsequent treatment in the vapour compartment of the degreaser. Provision is also made in the equipment for draining the liquid compartment from time to time in order to remove any build-up of soil sediments and so retain the high efficiency of the equipment.

The solvents used in this type of plant are the chlorinated hydrocarbons, the most commonly used being trichloroethylene. This solvent does not itself attack most metals but a violent reaction can occur between the hot solvent and finely divided light alloy metals, so precautions must be taken when handling these materials. Cases have been known where highly destructive explosions have resulted from the introduction into vapour degreasers of light alloy components carrying fine swarf or metal dusts. Special formulations of solvents containing additives to increase their stability have been developed, however, so these alloys can be safely treated. Care must also be taken to avoid a build-up of acidity in the solvent, since an acidic solvent may readily attack the metal articles being cleaned and, in serious cases of acidic build-up, the materials of construction of the plant itself. Wet work cannot be treated in a trichloroethylene plant, but if perchloroethylene is used as an alternative solvent wet work can be safely treated.

-13-

In addition, perchloroethylene has a higher boiling point (121°C) than tri-chloroethylene and this leads to improved efficiency of removal of hard greases, although at the expense of greater heating costs.

As an alternative to (or in a number of cases in combination with) solvent cleaning, soils and greases can be removed by chemical cleaning methods. Chemical cleaners can act on the soils in a number of different ways such as solubilising, emulsifying, saponifying and peptising. An alkali detergent-powder mixture is the most commonly used basis for chemical cleaners.

Alkali metal silicates, phosphates and carbonates are employed as hot aqueous solutions; the addition of surface active agents serves to lower surface tension so that the soiled work is more readily wetted by the clean-er, and promotes emulsification of oils and greases. The alkali salts them-selves have good detergent properties, causing saponification by reaction with fatty substances and promoting peptization which assists the retention of insoluble soils in suspension in the cleaner. Sodium metasilicate and trisodium phosphate are among the most commonly used of the alkali salts, but they are often fortified in purpose-formulated cleaners by additions of tripoly- or hexameta-phosphates, which chelate hardness salts in the make-up water and prevent precipitation of insoluble deposits on the work or plant.

Although caustic solutions are more efficient saponifiers than sili-cates, phosphates and carbonates they will react with many metals--notably the light metals and alloys--and they are considerably more difficult to rinse from the surface of the work after treatment. Some free caustic may, however, be incorporated in heavy-duty cleaners.

Finally, buffering agents may be added to the formulation of alkali cleaners.

Treatment by alkali cleaners may be effected by immersion in soak tanks, efficiency being improved by agitating either the liquid or the work, or by spray application from pressure jets. Thorough water rinsing must always be subsequently employed.

The efficiency of chemical cleaners can be much increased, and the dan-ger of chemical attack on the metal reduced or prevented, by electrolytic action. A polarising current of ~ 500 A/m^2 at an applied voltage of 3-12 V is used, the work being made either anodic or cathodic according to the metal concerned. Ferrous metals are anodically cleaned and copper-base mat-erials are treated cathodically; in many cases a brief reversal of polarity is employed prior to removal of the work from the cleaner so as to remove any electrodeposited smuts. The cleaning action of the process depends on the formation of gas bubbles on the surface of the work as a result of the discharge of hydrogen or oxygen gas at the metal surface. The bubbles of gas are formed at the metal surface beneath the soils and provide a mechan-ical removal action. In addition, cathodically produced alkali improves detergent action. Electrocleaning is not suitable for the treatment of tin, lead, zinc, aluminium or light alloys.

Both solvent and chemical cleaning may be assisted and their efficiency improved by employing ultrasonic agitation of the work while it is immersed in the cleaning liquid. A transducer built into the liquid tank induces ultrasonic vibrations in the immersed work and bubbles of gases or cavitation bubbles are produced at the work surface. When these bubbles either form or collapse, mechanical loosening or removal of the soils attached to the surface takes place thus improving the efficiency of the cleaning process.

Removing superficial corrosion

Superficial corrosion of metals occurs as a result of oxidation during processing (e.g. in hot-working processes and in heat-treatment processes) or through reaction with a corrosive environment during storage. Although the extent of this corrosion can be controlled and minimised by appropriate control during processing and by the use of temporary protective measures during storage it is extremely unlikely that it can be completely prevented. Any corrosion products on the metal surface must be completely removed before coatings are applied since their presence interferes with the application and/or the performance of the coating. Loose or brittle oxide films entrapped between the coating and the substrate produce regions of poor adhesion where breakdown of the coating can easily occur in service. Corroded areas may not be receptive to electrodeposition so that bare areas remain after plating, and the difference between the electrode potential of a corroded region and that of the rest of the substrate can produce local electrochemical action leading to enhanced corrosion in service.

The removal of unwanted processing or storage corrosion may, of course, be effected by mechanical means during machining, polishing or abrading, which are discussed in the next part of this chapter. Apart from the use of these methods, removal of corrosion is generally achieved by chemical immersion treatments known as pickling.

The pickling process is in many ways akin to the chemical cleaning processes already described--indeed, pickling is only a more aggressive form of chemical cleaning aimed not at greases or soils but at oxides and other more stable metal compounds. Removal may be achieved either by solution of the corrosion products in the pickle liquor or by their physical detachment from the metal surface when they are undermined by chemical attack on the substrate.

The aggressivity required of a pickle makes it necessary to move from the nearly neutral or mildly alkaline salts used as chemical cleaners to stronger acids or alkalis. The concentration and operating temperature are increased as the pickling duty moves from the removal of light tarnish stain to the removal of heavy oxidation and scaling. Once again specific formulations may involve the use of wetting agents to improve efficiency and speed of action, and inhibitors to reduce, or even completely prevent, attack on the clean metal beneath the corroded surface.

A different approach to pickling that is of particular benefit in the case of heavy, tough and adherent scales is the use of molten salt baths.

The removal action in this type of pickling may combine chemical attack on the scale by the molten salt with a shattering of the continuity of the scale by differential expansion from the underlying metal as a result of the themal shock of immersion in the molten bath. This method of pickling is finding increased application in a number of fields and may be of particular benefit as a way of combining descaling and heat-treatment in a single operation. However, the process requires special equipment and skilled operators, is costly and may be hazardous. In addition, because of the very nature of molten salt pickling the process cannot be employed where exposure to high temperatures will adversely affect the mechanical properties of the metal to be descaled. Molten solium hydroxide and molten sodium hydride (NaH) are frequently used for this purpose.

As with chemical cleaning, the action of pickling may be assisted by electrolytic action (using either anodic or cathodic polarisation of the work) or by the use of ultrasonic agitation.

The use of different types of pickling treatments for the various metals is summarised in Table 3.1.

Table 3.1 – **SUMMARY OF PICKLING METHODS FOR DIFFERENT METALS***

Metal	Soak cleaning	Immersion pickling	Electrolytic pickling	Salt-bath descaling
Iron or steel	Dilute acids used for removing light corrosion only. Pitting can occur with cast iron	Simple acid solutions used for removing rust or scale from plain carbon steels or cast irons. Stronger acid mixtures used for alloy steel. High-strength steels may suffer hydrogen embrittlement. Cast irons may become pitted	Anodic or cathodic treatment in acids used for steels especially prior to electroplating. Alkaline processes suitable for treating cast iron.	Mainly used for removing heavy scales from alloy steels and for removing siliceous scales from cast iron
Copper-base alloys	Dilute sulphuric acid used for removing light tarnish	Dilute mineral acids, often in mixtures or with addition of dichromate salts, used for removing heavier oxide scales	Mild cathodic alkali processes used for removal of light tarnish	Mainly used to remove very tough scales or adherent siliceous scales
Zinc and its alloys	Very dilute acids only used with short duration treatments		Not used	Not used
Tin and lead	Dilute acids used for removing light tarnish	Fluoboric acid solutions used for general pickling	Not used	Not used

Table 3.1 (cont'd.)

Metal	Soak cleaning	Immersion pickling	Electrolytic pickling	Salt-bath descaling
Aluminium and its alloys	Dilute acid or alkali solutions used for light etching only. Smut deposits removed by subsequent nitric acid dipping	Nitric/hydro-fluoric acid mixtures and hot chromic/sulphuric acid mixtures used for general pickling. Hydrofluoric acid or caustic alkali mixtures used for etching	Not used	Sodium hydride used for removing adherent siliceous scales
Magnesium and its alloys	Not often used	Chromic/hydro-fluoric, nitric, phosphoric, acetic and sulphuric acids all used in combinations for general pickling and etching	Not used	Not used
Nickel and its alloys	Not used	Sulphuric and hydrofluoric acids used for general pickling	Cathodic treatment in acids	Little used except for heat-resisting high-nickel alloys
Titanium	Not used	Sulphuric acid used for removing light scale. Fluoboric, hydrofluoric and nitric acids and mixtures used to remove heavier scales	Not used	Frequently used for removal of very heavy scale. With caustic salts treatment temperature must not exceed 480°C

*Based on data in *Finishing Handbook and Directory*, Sawell Publications Ltd (1970).

Controlling the physical nature of the surface

Primarily, the required condition of the surface of a metal to be coated is governed by the end-use of the finished product. Most coating processes can be applied equally well to cast, wrought, polished or roughened surfaces provided always that these surfaces have been thoroughly and scrupulously cleaned as indicated in the foregoing sections.

The one notable exception to this generalization is the case of coatings applied by metal spraying processes. The way in which sprayed metal coatings are built up is such that in order to achieve adequate adhesion between the sprayed coating and the substrate the surface of the latter needs to be roughened so as to provide a mechanical keying action to retain the coating during service. The degree of roughness and the angularity of the surface irregularities both markedly affect the adhesion, and it is also important to ensure that the roughened surface is free from contamination.

Pretreatment for metal spraying, therefore, is accomplished by grit-blasting, taking care to see that the range of grit sizes used is carefully controlled. Too fine a grit produces a surface with insufficient roughening for adequate adhesion; too coarse a grit produces an unacceptable degree of macro-roughening while probably still having insufficient micro-roughening to achieve the optimum coating adhesion. The actual range of grit sizes used depends on the materials of which the grit is composed, upon the metal that is to be treated and also, to a lesser degree, upon the air pressure supplied to the grit-blasting equipment. Chilled iron and alumina grits are the two materials most commonly used. In the interests of economy of materials it is usual to collect the grit and recycle it for further use; this is usually done by means of suction pipes placed adjacent to the treated surface during blasting or by carrying out the blasting operation in an enclosed cabinet from which the grit is collected and piped back to the blast nozzle (Figure 3.2). Care must always be taken to remove dirt and excessive fines produced when larger sizes of grit shatter in use.

The freshly abraded metal surfaces produced by grit-blasting tend to be chemically active and thin, air-formed oxide films are readily formed on them. For this reason operators should not handle the grit-blasted surfaces without using gloves and the sprayed metal coatings should be applied as rapidly as possible after grit blasting since any deterioration will adversely affect the performance of the coating. The delay between grit blasting and metal spraying should never be sufficient to allow visible deterioration of the surface to occur; the time limits to avoid this vary with the conditions under which the operations are being carried out. Specifications for metal spraying lay down maximum permissible delays between grit blasting and metal spraying; e.g. Defence Standard 03-3 (Protection of Aluminium Alloys by Sprayed Metal Coatings) allows a maximum of four hours under good workshop conditions and suggests that in on-site applications the delay should not exceed a few minutes.

Apart from its use as a pretreatment for metal spraying, the grit-blasting process may be used for materials subsequently coated by other methods. In these cases it is used to remove heavy scale from a metal prior to the employment of other pretreatment processes or to provide a surface with a controlled degree of roughness that may be required for decorative or frictional purposes. For achieving the finer grades of roughening--known as satin finishes--the process of vapour blasting may be used; this is essentially the same type of process as grit blasting except that very fine abrasives are used and are applied to the work by means of a pressure jet of water vapour.

Where processes other than metal spraying are used to apply metal coatings to cast, wrought (i.e. rolled, forged or as extruded) or machined surfaces it is only necessary to ensure that greases and soils or oxide films and scales are removed by using the appropriate cleaning techniques previously discussed. Three other classes of surface finishes may, however, be

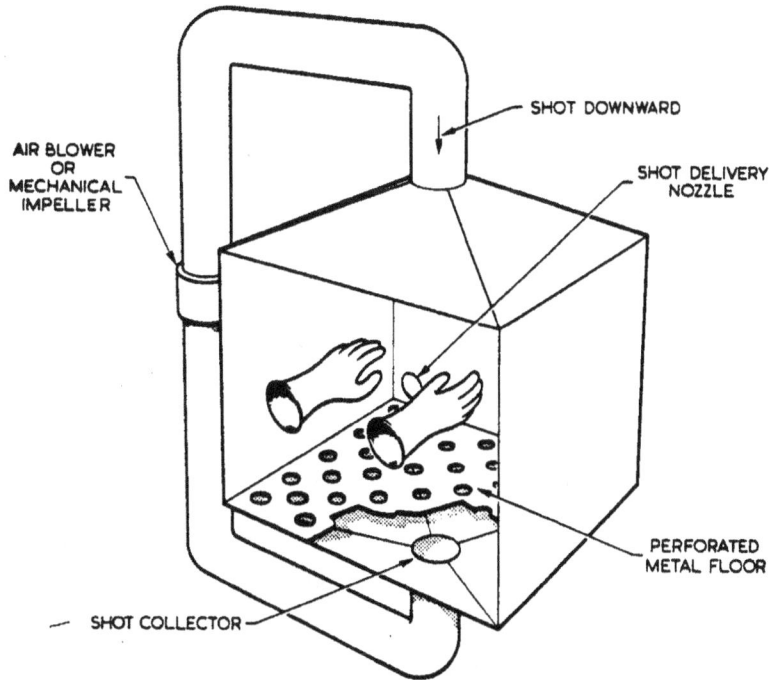

Figure 3.2. *Grit blasting cabinet*

required--abraded, polished and etched. These conditions are achieved in the following ways.

Abrading or grinding

Abrading consists of the controlled removal of metal from a surface by the application of grits of graded coarseness or by the use of rotary wire brushes. The grits may be cemented to paper, cloth or metal bands, strips or discs, and usually consist of tungsten carbide, alumina, diamond or siliceous materials supplied in a range of carefully controlled coarsenesses. The abrading process may be carried out by hand or by mechanical equipment and performed either in the dry state or lubricated with water or oils. Some degree of macrolevelling of the surface is effected, but a micro-roughened finish is produced that may be directionally or randomly oriented according to the way in which the process is applied. The pressure used to apply the abrasive to the work and the type and extent of lubrication used must be carefully controlled to avoid embedding particles of metal debris into the surface, where their presence could lead to the formation of defects in subsequently applied metal coatings.

Abrading or grinding may be used prior to polishing or as finishes in their own right. In the latter case they must always be followed by degreasing and/or cleaning treatments in order to remove soils and metal dusts before applying metal coating processes.

Polishing

Polishing is used to improve appearance, levelling, reflective properties or closeness of fit with mating components or to reduce friction between moving components. Mechanical, chemical and electrolytic methods may be used.

Mechanical polishing

This method may be considered as an extension of the abrading process at the finest end of the scale; metal removal is reduced and smoothing action accentuated. When applied in its final stages to produce the greatest lustre and smoothness it is known as buffing.

Very fine grades of emery, alumina or silicon carbide are used as polishing abrasives and are applied to the work by means of resilient felt or cotton mops or wheels. Tripoli and rouge may be used for obtaining the finest lustres. The abrasive may be fixed to the wheel with glue or retained in position by tallow or grease compounds used to lubricate the polishing process. Lubrication serves to assist metal flow (and hence smoothing) and to prevent gouging and embedding of abrasive particles in the metal surface. Localized heating of the surface during polishing, caused by friction, can also assist the polishing action.

Great care must be exercised when polishing the softer metals, in which excessive metal flow and smearing can occur. A particular example is zinc diecastings, where sub-surface porosity can occur beneath the casting skin; if excessive metal flow is allowed to occur the sound cast skin may be broken to reveal the porosity holes, with consequent bad appearance and the danger of entrapment of subsequent processing solutions.

After mechanical polishing very thorough degreasing and/or chemical cleaning must be carried out to ensure complete removal of particles of metal and abrasive and the removal of greases or waxes used for lubrication.

Mechanical polishing and burnishing can also be carried out by treating the work in rotating barrels or vibrating tubs. The articles to be polished are loaded in the containers together with ceramic or metal shapes or chips, polishing compounds, and water to act as a lubricant. Chemical buffering salts and wetting agents may also be added. The rubbing action between the work and the chips during rotation or vibration of the containers enables the polishing compound to act to remove metal from the surface of the work and so produce smoothing and brightening. Careful control of the components of the mix in the containers and the total loading and rate of rotation or vibration enable an optimum of polishing to be achieved without mechanical damage to the work or loss of detail in shaped parts.

Chemical polishing

The amount of true polishing that can be achieved by chemical immersion treatment is limited, and the process would perhaps be more accurately described as chemical brightening. Thus it is not possible to achieve a true mirror surface, though some degree of smoothing does occur and general brightness and reflectivity are improved.

The essence of the process consists of acid dissolution of the metal from the surface, the rate of attack being limited by controlling the rate of diffusion of soluble salts from the surface and the replenishment of free acid in the region. This is normally achieved by increasing the viscosity of the polishing solution and adjusting the formulation so that it contains large, complex molecules. Under slow rates of diffusion replenishment of fresh acid is slowest in deep recesses on the surface of the work and most rapid on asperities. Consequently, more metal is removed from the high spots of the article and a degree of micro-levelling may be achieved.

Most of the commercially available chemical polishing solutions rely on a chemical such as ortho-phosphoric acid to increase viscosity, the active reagent for metal dissolution being an oxidising acid such as nitric acid. Buffering agents and other salts to control dissolution rates may be included in the formulation. The process is usually operated at an elevated temperature, and work may be treated either individually or in batches contained in baskets constructed from materials resistant to the action of the polishing solution. Considerable quantities of toxic fumes are produced and must be efficiently extracted. The work must be rinsed very quickly and thoroughly after treatment since any polishing solution retained on the surface of the work after removal from the bath will continue to react with the metal until complete exhaustion occurs, thus leading to uneven results.

The only pretreatment required for chemical polishing is the complete removal of greases or other adherent soils to ensure that the surface of the metal is completely wetted by the polishing solution. However, the limited extent of the polishing action achieved makes it essential that a fairly high-quality surface finish be obtained before applying a chemical polishing process.

Some limited degree of regeneration of chemical polishing solutions can be achieved by the addition of carefully controlled quantities of the active reagents consumed in the process, but the build-up of metal salts in the solution has a maximum tolerance level after which polishing actions becomes progressively reduced. A further problem with chemical polishing is the active state of the metal surface after processing; this is revealed by the greater rapidity with which light tarnish develops, and in order to avoid deterioration from this cause it is necessary to apply any subsequent coating process immediately or to make use of temporary protectives.

Aluminium and copper and their alloys are the commonest metals treated by chemical polishing techniques; formulations are also available for the chemical polishing of silver.

Electropolishing

A characteristic of solutions used in chemical polishing is their high redox potential, which is due to the addition of powerful oxidizing agents such as nitric acid; during polishing these are cathodically reduced at a high rate with concomitant rapid anodic dissolution of the metal. On the other hand, in electropolishing the metal is made the anode of the cell in which the cathodic reaction occurs at another electrode--the cathode (an inert conductor such as platinum, stainless steel, carbon, etc.). Thus, whereas in chemical polishing the potential is controlled by the redox potential of the solution (limited to + 1.1 V versus SHE), much higher and more controllable potentials can be achieved in electropolishing by the e.m.f. supplied by an external source. The solutions used are far less aggressive than those used in chemical polishing and are frequently reducing acids, although again they are formulated so that they are viscous.

Although many of the theories to explain electropolishing invoke the concept of diffusion-controlled dissolution, Hoar (Nature, 165, 64, 1950) was one of the first to propose that crystallographic etching (without polishing) was suppressed by the formation of a thin compact solid film on the surface. Under these circumstances the anodic process is determined by the random appearance in the solid film at the metal/film interface of cation vacancies into which random metal cations can pass. This random dissolution gives a smooth micropolished surface.

Thus when a metal is made anodic in a suitable electrolyte, dissolution begins to occur. At low current densities dissolution is crystallographic and the surface becomes etched, but as the current density is increased a critical range is entered within which the diffusion layer is produced adjacent to the anode and reaches a maximum thickness.

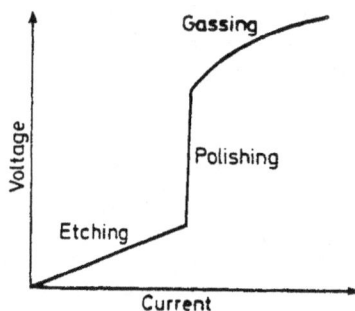

Figure 3.3. *Effect of voltage-current relationship on electropolishing*

The process then comes under diffusion control and in this range of operating conditions polishing action occurs, with true microlevelling of the surface so that mirror finishes can be obtained. Any further increase in current density leads to gassing at the anode and polishing action is lost. See Figure 3.3.

A wide range of solutions is available for electropolishing metals, each being specific to a particular metal or group of metals or alloys. Generally these solutions are much less concentrated and hence more safely handled than those employed for chemical polishing, and toxic fumes are not normally evolved during electropolishing processing. Electropolishing solutions tend to have a longer active life than chemical polishing solutions since in many cases metal taken into solution from the work is plated out on the cathode. Most of the ferrous and non-ferrous metals and alloys can be readily treated by commercially available electropolishing solutions.

As with chemical polishing, a clean and wettable surface is the only prerequisite of treatment by electropolishing techniques. However, whereas chemical polishing can be undertaken on batches of components, all items for electropolishing must be individually jigged or wired so as to provide the necessary electrical connections and this adds considerably to the operating costs of the process.

In chemical polishing, apart from the micro-smoothing, metal removal occurs uniformly over the whole of a shaped surface, whereas in electropolishing metal removal is less in very deep recesses and is minimal on the back of the article, which is shielded from the cathode. This feature of electropolishing can be exploited where it is desirable that metal removal should be limited in certain areas of shaped work.

Rinsing after electropolishing must be thorough but need not be so rapidly carried out as with chemical polishing, since the solutions used for electropolishing generally only attack the metal minimally in the absence of the polarizing current. Furthermore, electropolished surfaces tend to be somewhat more tarnish-resistant than chemically polished surfaces, so subsequent processing does not need to be carried out so rapidly.

Etching

Etching of metal surfaces may be chosen as a pretreatment process in order to produce satin-finish decorative effects, or to produce micro-roughening where a key is necessary to improve the adhesion of any subsequently applied coatings. Since the action is one of controlled roughening (or 'anti-levelling') it follows that the necessary degree of macro-smoothness of the contours of the work must be achieved by controlling the manufacturing processes or by using a grinding or part-polishing process before the etching stage is applied.

Etches always work by random dissolution of metal from the surface, as judged on a micro-scale, but their action is frequently highly selective on a micro-scale since individual grains of the microstructure of the metal may be attacked or inert according to their orientation. The action may be wholly chemical, chemical assisted by anodic electrochemical action, or wholly anodically electrochemical.

Both acidic and basic solutions may be employed in chemical etching processes according to the metal to be treated. Aluminium and its alloys are commonly etched in caustic-based solutions, to which may be added buffering and wetting agents, sequestrants, and a range of salts to control the severity of the etch and to complex the aluminium ions. Alternatively, acid-based solutions may sometimes be used. Copper-base alloys and ferrous materials are normally etched in solutions of the oxidizing or mineral acids that are also employed for pickling processes, but the concentration of acid is generally less than that used in pickling; metal salts are commonly added to the etches to produce a partial inhibition of metal removal or to modify the dissolution process so that it becomes more selective on certain features of the metallurgical structure of the metal.

Anodic etching of ferrous materials is normally carried out in a sulfuric acid solution, the concentration required being inversely proportional to the severity of the etch required. The anodic current density and the treatment time are increased in order to obtain deeper etching effects, and gassing occurs at the anodes. Copper may be anodically etched at high current densities in a solution of mixed chlorides; the process is widely used for halftone printing plates.

Etching is widely used (particularly in printing, engraving, and electronic applications) for the selective removal of metal from closely defined areas and to produce raised or recessed patterns. In order to achieve these effects the metal surface is first coated with a wax, or other material resistant to the etching solution, in those areas where removal of the metal is not required; etching is then confined to the uncoated portions of the metal surface. Finally, the wax coating is removed by solvents after etching and rinsing have been completed.

In order to obtain uniform etching, metals to be treated must be free from scale, soils and greases and must be completely wettable by the etching solution. After etching, thorough rinsing must be carried out; as with chemical polishing, this rinsing should be done as rapidly as possible in order to prevent etchant retained on the surface from continuing to react with the metal in localized regions.

Section 4

Methods of Applying the Coating

Many methods are available for forming a coating of a corrosion-resistant material on a corrosion-prone substrate metal. A brief description will be given of these different methods.

Cathodic Sputtering. A partial vacuum is required. The part to be coated is attached to the anode and a low pressure of an inert gas such as argon is admitted to the system. The positively charged gas ions are attracted to the cathode. The collision of the gas ion dislodges atoms from the cathode, which are in turn attracted to the anode and coat the part. One of the major advantages of cathode sputtering is that non-conducting as well as conducting substrates can be coated. Major disadvantages include the heating of the substrate and low deposition rates.

Diffusion Coating. This method requires a preliminary coating step which is followed by thermal treatment and diffusion of the coating metal into the substrate. A commercial material known as galvannealed steel is made by coating steel with zinc followed by heat treatment and the formation of an iron-zinc intermetallic coating by diffusion.

Electrophoretic Deposition. Finely divided materials suspended in an electrolyte develop a charge resulting from asymmetry in the charge distribution caused by the selective adsorption. Coatings may be formed by immersing the substrate metal in the electrolyte and applying a potential. If the particles have a negative charge, they will be deposited on the anode and if they have a positive charge, they will be deposited on the cathode. Commercial applications of this method are limited.

Electroplating. One of the more versatile methods for forming a metallic coating is by electrodeposition, i.e., making the metal to be coated the cathode in an electrolytic cell and applying a potential between the cathode on which the plating occurs and the anode which may be the same metal or an inert material such as graphite. The method is applicable to all metals which can be electrolytically reduced from the ionic state to the metallic state when present in an electrolyte. Certain metals such as aluminum, titanium, sodium, magnesium, and calcium cannot be electrodeposited from aqueous solution because the competing cathodic reaction, $2H^+ + 2e^- = H_2$, is strongly favored thermodynamically and it occurs in preference to the reduction of the metal ion. These metals can be electrodeposited from conducting organic solutions in which the H^+ ion concentration is negligible.

The mass of electrodeposited coatings can be accurately controlled because the amount deposited is a function of the number of coulombs passed. If the current efficiency for the metal deposition process is less than 100%, as for example when some hydrogen is also formed, it is necessary to know the current efficiency before the deposit mass may be calculated. Of equal importance to the mass deposited is knowledge of the distribution of the deposit. The ability of an electroplating bath to deposit uniform thicknesses at all sites on the cathode is characterized by the term "throwing power." A bath with good throwing power has a greater tendency to deposit the metal in uniform thickness than one with poor throwing power. Quantitative measurements of throwing power may be obtained in a Hull cell in which the cathode is oriented in a non-parallel way with the anode.

The morphology of the electrodeposit may be controlled by the addition of agents to the plating bath. Bright electrodeposits of nickel may be obtained, for example, by the addition of surface active agents such as sulfonates to the bath. Nickel plating baths that have the ability to fill in minor scratches and depressions are obtained by the use of small amounts of coumarin in the bath.

Table 4-1 gives some information about metals that are commonly electroplated. Many alloys may be electrodeposited. Some of the more important include copper-zinc, copper-tin, lead-tin, cobalt-tin, nickel-cobalt, nickel-iron, and nickel-tin. The copper-zinc alloys are used to coat steel wire used in tire cord; lead-tin alloys are known as terne plate and have many corrosion-resistant applications; nickel-tin alloys have been proposed as a substitute for gold in protecting electronic components from corrosion.

TABLE 4-1

General Information about Electrodeposited Metal Coatings

Metal	U.S. Consumption as Plated Metal in lbs.	Plating Bath Type	Applications
Chromium	50,000,000	Sulfuric Acid	Decorative coating, wear resistant coating, generally applied over nickel
Nickel	48,000,000	Sulfate-chloride	Undercoat for chromium on automobile parts and plumbing fixtures
Copper	35,000,000	Sulfate; cyanide	Rotogravure rolls, undercoat for nickel, as conductor in electronic equipment
Zinc	40,000,000	Cyanide; chloride; alkaline zincate	Electrogalvanized steel
Silver	32,000,000	Cyanide	Decorative coating; leads for electrical conductors
Gold	2,000,000	Cyanide	Decorative coating; protection of electronic components

Explosion Bonding as its name implies involves the development of a bond between two metals by the exertion of a strong force that compresses the two metals sufficiently to develop a strong interfacial interaction.

Flame Spraying. A fine metal powder or a wire is passed through a flame whose temperature is sufficient to melt the metal and maintain it in the molten condition until it strikes the part to be coated. The method is used with aluminum and the density of the resulting deposit is lower than that of pure aluminum suggesting the presence of voids in the coating.

Fusion Bonding. Low melting materials such as tin, lead, zinc, and aluminum may be applied as a coating by cementing the metal as a powder to the surface and then heating the part to a temperature above the melting point of the coating metal.

Gas Plating. Some metal compounds may be decomposed by heat to form the metal. Outstanding examples are metal carbonyls, metal halides, and metal methyl compounds. Nickel deposits may be obtained by the thermal decomposition of nickel carbonyl. Production processes for titanium and zirconium are based on the formation of iodides which are transported in the gas phase to a hot wire where the iodide is decomposed to form the metal.

Hot Dipping. Large tonnages of steel are coated with zinc by immersion of the metal as a continuous sheet into a molten bath of zinc to form galvanized steel. The establishment of a good bond at the zinc-steel interface requires the presence of a small amount of aluminum in the bath. The thickness of the zinc coating is controlled by rigid control of the temperature of the galvanizing bath, the speed of transit through the bath, the temperature of the steel sheet before it enters the bath, and the use of air jets which exert a wiping action on the molten zinc as the sheet emerges from the bath. Tinplate was formerly manufactured by hot dipping but practically all commercial tinplate is now made by electrodeposition because of the ability to control the thickness of the tin.

Immersion Plating. This method also known as electroless plating or chemical plating is based on the formation of metal coatings resulting from autocatalytic chemical reduction of metal ions from solution. The solution must contain a reducing agent and the surface on which the deposit occurs must be catalytically active and remain catalytically active as deposition proceeds. Metals commonly plated by this technique include nickel, copper, silver, cobalt, and palladium. The silvering of mirrors falls in this classification. Typical reducing agents include hypophosphite, amine boranes, formaldehyde, borohydride, and hydrazine. Electroless nickel deposits formed with hypophosphite as a reducing agent contain phosphorus and many of the properties are determined by this alloying constituent.

Metal Cladding. Clad and mechanically formed coatings are composites of two or more metals that have been joined together in solid state without the use of intermediate binders. Good bonding requires that the mating surfaces be free of oxides, grease, moisture, and other contaminants before the cladding step. The most common method of manufacture involves roll bonding in which there is a 50-75% reduction in thickness. Bonding may be carried out with or without prior heating of the metals.

Plasma Spraying.. This method is analogous to flame spraying except that the energy is applied by rf heating rather than by a flame.

Vacuum and Vapor Decomposition. This technique receives its greatest use in the formation of metallic coatings on non-conductive substrates. Rhodium coatings on mirrors and aluminum coatings on plastics are the more common vacuum deposited coatings.

Section 5

Evaluation of Surface Character Prior
to Application of the Coating

Metallic coatings are only effective if there is a good bond between the substrate metal and the coating. Any foreign material that is not purposely applied to improve the bonding will reduce strength of the bond. Common foreign materials include dirt and dust, corrosion products such as iron oxide scale, grease and oil, abrasive particles, marking materials, and corrosion inhibitors. The methods for removing these materials have been covered in Section 3. The purpose of this section is to outline some of the procedures used to appraise the character of the surface. The four characteristics that will be discussed include: cleanliness, surface chemistry, topography, and surface reactivity.

Cleanliness. This term generally applies to a determination that the metal surface is free of organic materials. A metal that is free of organic contaminants has a high surface energy and is completely wetted by water. Contaminated surfaces cause water to bead up in droplets having a finite contact angle with the metal. Measurements of this contact angle give qualitative information about the cleanliness and how the cleanliness changes with time.

Organic contaminants are present in the air and as impurities in most aqueous solutions. They are ubiquitous and are very hard to avoid. Clean metals exposed in air rapidly pick up organic contaminants and the metal is no longer clean after a short period of time. Some data on the effect of storage conditions on the contamination of clean nichrome are given below:

Effect of Storage Condition on the Variation
of Contact Angle of Water on Nichrome

Storage environment	Contact angle of water on Nichrome after 16 hr of storage, deg
Initial (after cleaning)	6
Laboratory air	35
Stoppered glass bottle	30
Empty aluminum desiccator	18
Desiccator with activated charcoal	16
Desiccator with activated alumina	12
Desiccator with aluminum shot	6

Data taken from M. L. White, "Clean Surfaces," G. Goldfinger, Ed., Marcel Dekker, New York, 1970, pp. 361-373.

Additional data that show the rate of surface change as a function of time of storage are given in the following figure taken from R. E. Pike, "Corrosion Control by Organic Coatings," H. Leidheiser, Jr., Ed., Science Press, Princeton, N.J., 1979, pp. 259-276.

SURFACE ENERGY AS A FUNCTION OF CLEANING

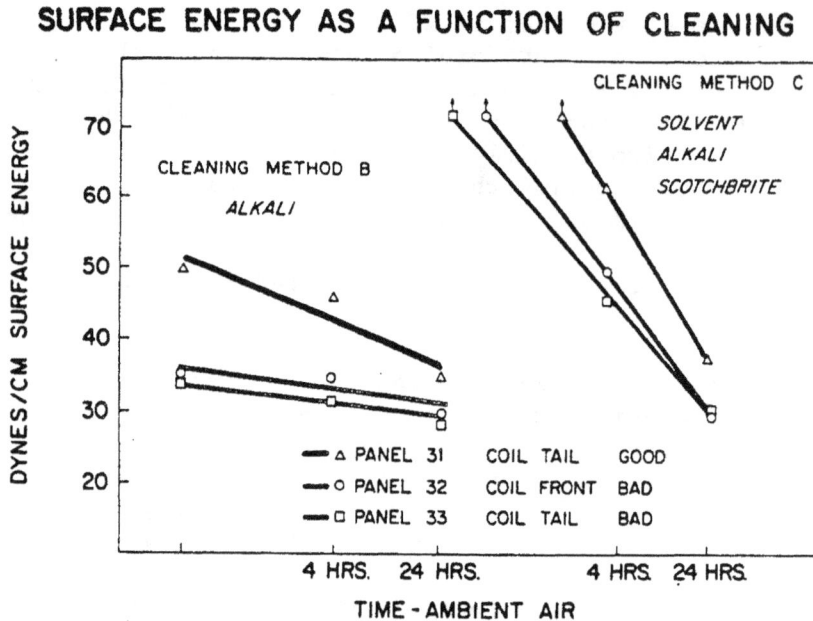

Surface energy of steel surfaces as a function of cleaning method, steel quality, and time after cleaning.

Surface Chemistry. The chemical composition of the very outermost layers of atoms on the metal surface play a dominant role in determing the activity of the metal. Clean metal surfaces are generally covered with a thin oxide which develops on exposure to air. The metal surface may also contain impurities that remain on the surface after a processing step. Common contaminants include organic compounds adsorbed from the atmosphere, calcium and fluorine picked up from exposure to water, chlorine picked up from water or from chlorinated cleaning solutions, carbon picked up from reaction with carbon dioxide in the air, alkali metals such as sodium or potassium picked up from water solutions, etc.

These contaminants and any others with atomic numbers greater than that of lithium can be detected on the surface by two different techniques, both of which require that the sample be exposed to a very high vacuum. Both techniques depend on irradiating the sample with electrons (Auger Spectroscopy) or with X-rays (X-ray Photoelectron Spectroscopy) and determining the energy of electrons that are emitted by the sample during a deexcitation step. The electron energy is a characteristic of the element giving rise to the electrons. A spectrum of the electron energy of the emitted electrons allows one to make a qualitative analysis of the composition of the surface.

Topography. Two techniques, profilometry and electron microscopy, are commonly used. The most widely used profile measuring instrument is the electromechanical stylus or tracer instrument. In these instruments a small (1-10 μm) diamond stylus is mechanically scanned across the measured surface. Stylus forces are of the order of tens to hundreds of milligrams. The vertical deflection of the stylus activates an electromechanical transducer, somewhat similar to a phonograph pickup, whose electrical output is amplified and either recorded on chart paper or measured by a special averaging type electrical meter. The meter records a one dimensional characterization of the surface: the arithmetic average deviation of the measured surface from an ideally smooth surface (some older U.S. instruments measure RMS roughness). Transducers for converting the mechanical displacements of a stylus into electrical currents or voltages are of two basic kinds: (1) displacement, carrier modulating devices in which the amplitude of an alternating current of high frequency is modulated by the displacement of the stylus, regardless of how long the stylus remains in a given position and (2) motion, devices in which the electrical signal is generated according to the motion of the stylus as it is displaced from one position to another. The displacement responding types produce a signal which is directly related to the profile, whereas the motion type employ an electrical integration process, limiting severely the reliability of the recorded profile. Both types provide reliable roughness numbers as indicated by the averaging meter.

The vertical resolution of a high quality stylus instrument is 0.01 micrometers or less. Horizontal resolution is determined primarily by the stylus radius (1-10 μm). Since the stylus damages softer surfaces, the instrument's chief application is to hard surfaces. The electrical output of the stylus instrument may easily be digitized so the profile can be stored in the memory of a minicomputer and subsequentally analyzed. The scanning electron microscope is a flying-spot type of instrument with resolution intermediate between that of the transmission electron microscope and the optical microscope.

Contrast in the scanning electron microscope is affected by many factors. The number of secondary electrons depends strongly on the tilt of the region being probed and even more strongly on the presence of sharp edges. Electrical fields at the specimen surface, arising from the potential of adjacent electrodes, enhance emission at protrusions. The efficiency of secondary-electron production depends on the particular element present where the beam strikes the surface, the crystallographic surfaces exposed, the work function and the presence of adsorbed layers. The contrast gradient depends on the gain of the amplifier. For all these reasons one can be very badly misled in interpreting edges and bright and dark areas as topographic information. Stereoscopic pairs are essential, even for good qualitative interpretation of micrographs. Quantitative information can be obtained by painstaking analysis of stereo pairs of micrographs.

Surface Reactivity. In some cases it is important to know whether the cleaned metal has high reactivity. This reactivity may be important in terms of appraising the nature of the substrate metal or in appraising the likelihood of a good bond between the substrate metal and the metallic coating. One of the techniques that is useful is dependent on the determination of the rate of dissolution of the metal in an electrolyte while being subjected to a changing applied potential. The resulting curve, which is a plot of the potential of the metal vs. the current density flowing through the metal interface is known as an anodic polarization curve. A high value of the critical current density for passivity and a higher value of the current at a range of potentials is indicative of a higher reactivity. An example of two steel panels with different reactivity characteristics in such a test is shown in the figure below.

ANODIC POLARIZATION CURVES

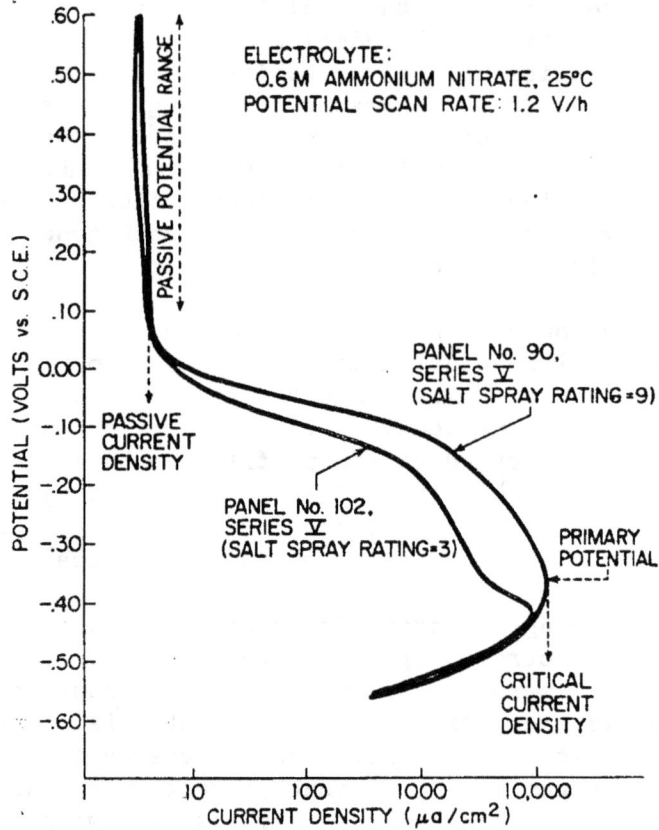

ELECTROLYTE:
 0.6 M AMMONIUM NITRATE, 25°C
POTENTIAL SCAN RATE: 1.2 V/h

PASSIVE POTENTIAL RANGE

POTENTIAL (VOLTS vs. S.C.E.)

.60
.50
.40
.30
.20
.10
0.00
-.10
-.20
-.30
-.40
-.50
-.60

PANEL No. 90,
SERIES $\overline{V}$
(SALT SPRAY RATING = 9)

PASSIVE
CURRENT
DENSITY

PANEL No. 102,
SERIES $\overline{V}$
(SALT SPRAY RATING = 3)

PRIMARY
POTENTIAL

CRITICAL
CURRENT
DENSITY

CURRENT DENSITY ($\mu a/cm^2$)

1   10   100   1000   10,000

Section 6

Methods for Measuring Important Chemical and Physical
Properties of Metallic Coatings

Much of the information contained in this section has been abstracted
from the book, "Properties of Electrodeposits. Their Measurement and Significance," R. Sard, H. Leidheiser, Jr., and F. Ogburn, Eds., The Electrochemical
Society, 1975, 430 pp.

The properties that will be considered are:

Adhesion
Corrosion Properties
Density
Electrical Resistivity
Interdiffusion
Magnetic Properties
Mechanical Properties
Porosity
Solderability·
Thickness
Topography
Wear and Contact Resistance

Other important properties, whose measurement is rather straightforward,
such as crystal size and crystal orientation, will not be treated here. These
characteristics are important in hot dipped zinc coatings, for example, because products with a high degree of (0001) preferred orientation and small
grain size show better paint adherence when the painted galvanized steel is
deformed during industrial forming operations.

Adherence

Qualitative Mechanical Tests

Tests	Principle and Comments
Bending and Twisting	Most commonly used, especially for thin deposits but also for thick deposits. Based upon the idea that the differ- ent lengthening of substrate and coating upon stretching side by side results into forces tending to separate the two. Variations of the method involve variable radii, re- peated flexing, various set angles (90° or 180°) of bend, wrapping and twisting. Bending radius, thicknesses of both substrate and coating all affect the adhesion strength.

Tests	Principle and Comments
Cont'd.	Not applicable to brittle or hard coatings, because cracks generally develop. Ductile coatings may reduce the applied stress by plastic flow, consequently, poorly adherent coatings may not be detached.
Impact or Hammering	These tests are drastic in nature and combine effects of deformation, shock, impact, local heating and fatigue. Based upon the repeated hammering or impacting with a single tool or the use of grits or shots propelled at the surface. The number of blows (1500-1600 per min) to remove the coating is a measure of adhesion. Exfoliations or blisters in and around indentations are evidence of poor adhesion. Soft and ductile coatings cannot, generally, be studied. The test can be made semi-quantitative by utilizing weights falling from various distances to strike a blow to the sample.
Burnishing, Buffing and Abrasion	A repeated rubbing of the surface with a tumbling medium or tool results in lifting in the form of blisters where adhesion is poor. Pressure, surface area of contact, area burnished, nature of the burnishing tool and speed of operation affect the final value of adhesion. This test combines deformation and fatigue. Can be applied locally to most production parts and is considered non-destructive where adhesion proves acceptable. No standardization. Thick deposits cannot generally be evaluated satisfactorily.
Heating and Quenching	Different thermal expansion between the coating and substrate or the pressure from gas or liquid vaporized from the substrate or interface causes disruption at the interface. However, unless the coating can be peeled or lifted from the substrate, the blisters are not indicative of poor adhesion. If the coating and substrate are sensitive to oxidation, these should be heated in an inert or reducing atmosphere or a suitable liquid. Diffusion and alloy formation accelerated by the heating may result in reduced or improved bonding. Time and temperature of heating varies with both the substrate and the coating. ASTM has listed the necessary heating temperatures for a variety of substrates and coating. Independent of the geometry of the part. Heating is sometimes followed by quenching to produce blisters. May be regarded non-destructive in cases of good adhesion.
Thermal Cycling	The coated part is subjected to abrupt temperature changes, typically from -40°C to +105°C. If the adhesion is poor, lifting of the electrodeposit occurs. Commonly used for electroplated thermoplastic materials.

Tests	Principle and Comments

Scribing or Scratch

Scratches are made through the coating in a variety of parallel or intersecting patterns. Observation of the lifting or peeling of the deposit between the scratches is the basis of evaluation and this can be done by using a pressure sensitive tape. The tests are dependent upon many factors including the operator's experience, distance between scratches, thickness of the coating. Thick coatings are not suitable unless a chisel or sharp knife is used to pry the interface in which case it becomes a chisel test (see below).

Chisel

Most severe of all adhesion tests. Applicable to thick coatings, but is limited by the thickness of the deposits and the toughness of available chisel tools. Soft and thin coatings are not suitable.

Grind Wheel

Involves holding the plated article against a rough emery wheel so that the wheel cuts through the coating in an irregular fashion. This method has been standardized by taking into consideration the wheel dimensions, speed, etc. Not suitable for thin or soft coatings. A similar test based on sawing has also been used.

File

A plated specimen is sawed and subsequently subjected to a coarse mill file across the sawed edge from the substrate toward the coating so as to raise it using an angle of 45° to the coating surface. Lifting or peeling is indicative of poor adhesion; thin or soft coatings are not amenable to this technique.

Cupping and Indentation

These tests are the outgrowth of Erichson Cup Test and metallurgical hardness testing. Dependent upon the deformation of both the deposit and the substrate in the form of a flanged cup or depression using a plunger. Peeling or flaking of the deposits is a measure of adhesion strength. Most satisfactory for harder or brittle coatings such as chromium and hard nickel; ductile coatings which deform easily are not amenable to such tests. Thickness of the deposit may affect the results.

Push-out

In this test, a blind hole is drilled from the rear of the coated part and a hardened punch is applied at a uniform rate to push out a "button" sample. Microscopic examination of the button and the periphery of the crater in the substrate is used as an evaluation of the adhesion strength; exfoliation or peeling of the film indicates poor adhesion. Suitable for a variety of basis metals, ductile or non-ductile, hard and brittle deposits. Apparently, a minimum thickness of the deposit is required; and soft, very ductile and thin deposits are not suitable.

Tests	Principle and Comments
Push-in	This is a variation of the push-out test. Essentially, small hemispherical indentations are pushed in to a predetermined depth on coated parts. In case of satisfactory adhesion, it is a nondestructive test. Visual examination can easily detect cases of poor adhesion.
Scotch Tape or Adhesive Tape	Essentially, it is a "go", "no-go" test. A pressure sensitive tape is pressed onto the coated part and then rapidly stripped. The extent of removal of the coating is a measure of adhesion. Very commonly used for thin films but not so popular among workers in electrodeposits.

Appropriateness of Qualitative Mechanical Tests for Various Electrodeposited Coatings

Adhesion Test	Ag	Au	Cd	Cr	Cu	Ni	Ni/Cr	Pb and Pb/Sn	Sn and Sn/Pb	Zn
Bend					*	*	*			
Burnish	*	*	*		*	*	*	*	*	*
Chisel				*	*	*				
Draw	*	*	*	*	*	*	*	*	*	*
File					*	*	*			
Grind & Saw				*		*	*			
Heat/Quench	*			*	*	*	*	*	*	
Impact				*	*	*	*		*	
Peel	*	*	*		*	*		*	*	*
Push				*		*	*			
Scribe	*	*	*		*	*			*	*

Quantitative Adhesion Tests

Tensile
Pull
A force is applied perpendicular to the coating/substrate interface and the amount of force per unit area necessary to disrupt the interface is a measure of the adhesion strength.

Disadvantage: Solders, adhesives or electroformed grips must be used to make contact with surface of coating. These foreign materials may affect the test results.

Shear Test
A cylindrical rod is coated with separate rings of the coating of predetermined width. The rod is forced through a hardened steel die having a hole larger in diameter than the rod but less than that of the rod plus the coating. The coating is detached and the bond strength, A, is determined from the formula, $A = W/\pi dt$, where d is the diameter of the rod, t is the width of the coating, and W is the force required for detaching.

Disadvantage: Equipment required. Not always easy to adopt the test to a particular system.

Peel Test
Requires that some part of the coating be de-adhered so that it can be gripped and the peel test performed.

Disadvantage: Results cannot be compared with other types of adherence tests.

Knife and
Scribing
Tests
Two categories of tests are used. In one case, the force required or the work done in detaching a coating by drawing a loaded knife or rounded point across it is measured; and in the second case, load is gradually increased on a pointer which is dragged across the coating and the critical load required for a clear track is recorded.

Blister
Method
A fluid-gas or liquid is injected beneath the coating at the coating/substrate boundary and the hydrostatic pressure is increased until the coating begins to detach from the interface.

Disadvantage: Not always easy to apply in practice.

Corrosion Properties

Several laboratory tests that appraise the protective properties of metallic coatings will be described.

Service Exposure. Adequate durability of a coated part in service is the only proof of satisfactory performance. The disadvantages of service exposure include (a) exposure durations are unacceptably long, (b) exposure conditions are variable, and (c) the number of parts that can be exposed is limited to a value about equal to the number of assemblies of which the part is a member.

Outdoor Exposure Tests. Sites are classified as rural, industrial or marine.

Neutral Salt Spray. Panels are placed in box into which a 5% solution of sodium chloride is aspirated with air. Temperature, pH, fallout rate of spray droplets and other important criteria are covered in ASTM B117-73.

Acetic Acid Salt Spray. Acetic acid is added to sodium chloride to yield a pH of 3.1. This test is rarely employed.

SO_2 Test. The coated panel is placed into a noncondensing humidity chamber containing about 1% SO_2. An exposure time of aout 24 hours is employed. Test is rarely used in the United States.

CASS and Corrodkote Tests. CASS stands for copper-accelerated acetic acid salt spray test and is used primarily for testing decorative chromium electrodeposits. The Corrodkote test requires application to the specimen surface of a paste containing copper nitrate and ferric and ammonium chlorides in a clay-water matrix, followed by exposure to a high humidity noncondensing atmosphere. The test requires 16-20 hours.

FACT Test. The Ford Anodized Aluminum Corrosion Test (FACT) is used only on anodized aluminum and is an electrolytic corrosion test. The specimen is made cathodic for 3 minutes in CASS solution. The galvanostatically impressed voltage decreases from an initial maximum of about 35 volts as the current increases with the growth of corrosion sites. The value of 750 volt seconds on an area of about 0.08 cm^2 is a minimum for acceptable performance.

EC Test. The electrolytic corrosion (EC) test was developed for use with copper-nickel-chromium and nickel-chromium electrodeposits. The plated surface is made anodic for 1 minute in a solution containing nitric acid, sodium nitrate and sodium chloride. The impressed potential is maintained constant compared to a reference electrode until the current density achieves a predetermined value; thereafter the current density is maintained constant by reducing the potential as required.

Density

The density is calculated from direct measurements of the mass and volume of the specimen. For coatings, this presents some problems because of the small mass and volume usually available.

Most of the measurements of density of metallic coatings have been made by hydrostatic weighing. The sample is weighed in air by conventional means. The apparent mass in water is determined by suspending the object in water from one arm of a balance by a fine wire. The density, $D = M/V$ may then be calculated from the two equations:

$$\text{weight in air} = M - \rho_A V$$

$$\text{weight in liquid} = M - \rho_L V$$

where ρ_A and ρ_L are the densities of air and liquid.

Electrical Resistivity

Two techniques are generally used, the two-point probe method and the four-point probe method.

In the two-point probe method, a long narrow conductor with a large pad at each end is fabricated either by vapor deposition or other technique. The resistance between the two pads is then directly measured using an ohmmeter or by measuring the current and the voltage drop between the pads. The basic advantage of the two-probe method is its simplicity but there are sources of error in the contact resistance between the probes and the pads.

In the four-probe method, the four probes are placed in a straight line; the two outside probes are current leads and the inside two are used for measuring the voltage drop. The measuring circuit is usually a version of the Kelvin bridge. The four-point probe has two advantages. Samples can be studied without any further processing and the method has a high degree of accuracy.

Interdiffusion. When a metal is coated on another metal there is always the possibility of interdiffusion between the two different phases. This interdiffusion may result in the formation of intermetallic compounds with different physical properties or a solid solution may be formed. It is often important to determine the degree of interdiffusion. The following techniques have been applied.

Electron probe microanalysis

Radioactive isotopes

X-ray fluorescence

Sectioning followed by metallographic study

Hardness measurements across the cross section

Magnetic Properties

The most fundamental macroscopic magnetic property of a magnetic material is the relationship between its magnetization, M (magnetic moment per unit volume), and the magnetizing field, H. Although there are conditions and field ranges where M is proportional to H, all materials saturate for sufficiently large magnetic field. Many show hysteresis, that is the value of M at a value of H can vary depending on the previous history of the sample.

The important techniques for determining the magnetic properties are:

> Alternating field magnetometer
>
> Vibrating sample magnetometer
>
> Force balance magnetometer
>
> Torque balance magnetometer
>
> Kerr Magneto-optic apparatus
>
> Four contact probe
>
> Microwave resonance

Mechanical Properties

The two important mechanical properties are tensile strength and hardness. The former can be measured by conventional tensile tests adopted for a thin film or by separating a portion of the substrate from the coating and applying a hydrostatic force causing the coating to bulge.

Thus, the most useful test which can be applied without separating the coating from the substrate is one in which the hardness is measured. Hardness is not really a property at all, but actually is a measure of a number of properties which interact in different ways for various materials and testing methods. Hardness is a measure of the resistance of the material to deformation. Permanent deformation is determined by yield strength, plastic flow and strain hardening.

The Vickers hardness method utilizes a diamond-shaped indenter which may be attached to an ordinary microscope for the indentation. The size of the deformation spot as a function of the applied load is used to calculate the hardness. The details of the test are covered in ASTM Standard E384.

Porosity

The most common test for detecting porosity of metallic coatings is to expose the coated metal to a fluid which penetrates the pores and corrodes the substrate. The corrosion product spills out of the pore. In the simplest tests sufficient product is formed to be visible to the eye. Usually the substrate is a corroding anode with the cathodic reaction occurring on the walls of the pore or on the external surface of the coating. A liquid reagent allows the use of an external cathode and an applied potential to modify or accelerate the reaction, but with increased probability of attacking the coating. A print of the pore pattern can be made by absorbing the product on moist paper pressed against the surface. This pattern may be visible at once or on subsequent development.

The conditions for a pore corrosion test are: (a) effective penetration of pores by the fluid; (b) generation of a sufficiently visible corrosion product; (c) while the product from the smallest pore should spread enough to be visible, that from the largest pore should not produce general stain.

Solderability

A brief description of the four basic tests will be given.

Capillary Penetration. A capillary space is provided between two flat metal sheets. The joint surfaces of the sheets are fluxed, preheated, and brought into contact with a molten solder bath for a specific time interval. After the sheets are removed from the solder and cooled, the capillary rise is measured on the wetted surface. Data from this type of test often show anomalous results due to nonuniform oxidation of metal surfaces or changes in the activity of fluxes during preheating.

Area of Spread Tests. Fixed volumes of solder and flux are placed on test coupons which are heated to a specific temperature. During heating, the solder melts and spreads over the surface. The extent of the solder spread pattern is influenced by the type of flux, metal surface condition and solder composition. The area of spread is determined after cooling.

Dip Tests. These tests are one of the easiest forms of solderability testing to apply since the specimen to be tested is immersed in solder and the degree of coverage by the solder is noted as an index of solderability.

Wetting Time Test. The apparatus for performing the test has three basic parts: a load-measuring system and read-out circuit, a motor drive for moving the solder bath vertically, and a solder heating and temperature control system. A test piece is suspended from a transducer-monitored spring system and the solder bath is raised until the specimen is immersed to a predetermined depth. Buoyancy from the displaced solder provides a transducer generated signal which is fed to an oscilloscope or high speed pen recorder. This condition is maintained until the solder wets the test specimen. The solder meniscus then rises, producing a downward force on the specimen, and a signal of opposite polarity is developed by the transducer. These first and second stages permit measurement of the time to start wetting and the rate at which wetting occurs. Eventually equilibrium is reached and the magnitude of the signal measures the ultimate extent of wetting.

Thickness

Measurements of thickness fall into two categories: destructive and non-destructive. The destructive methods are the following:

Coulometric. The coating is dissolved anodically at a constant current density and the thickness is calculated from the current density, area and density of the coating. Commercial instruments are available.

Stripping and Chemical Analysis. The mass of coating material may be determined by several methods: (1) The specimen may be weighed before and after plating. (2) The specimen may be weighed before and after dissolving the coating without attack of the substrate. The extent of dissolution of the substrate can be assessed by analysis of the solution or by running a blank on an unplated specimen. (3) The coating material may be weighed after dissolving the substrate in a suitable reagent. (4) The coating material may be dissolved along with part or all of the substrate and then its mass determined by chemical analysis of the solution.

Microscopic Methods. The coated metal is cross sectioned and the thickness is measured directly.

Step Height Methods. This method is adaptable to coating thickness measurements if part of the coating can be removed so as to leave a step between the coating and the substrate. Measurement techniques include interferometry, stylus tracing, micrometers and dial gages.

The non-destructive methods include the following:

Magnetic Gage. If the coating material and/or the substrate material is magnetic, the measured magnetic property is a function of the coating thickness and the instrument can be calibrated in thickness units.

Eddy Current Gage. A probe contains a coil carrying a high frequency current. When the probe is positioned on a test specimen, the electromagnetic field of the coil induces eddy currents in the specimen. The frequency of the coil current, between 100 kHz and 12 MHz, is chosen so that the eddy currents penetrate the plated coating and extend into the basis metal. The magnetic field of the eddy currents changes the impedance in the probe coil, and this change is measured by a suitable circuit. If the electrical conductivity of the coating differs from that of the substrate, the magnitude of the eddy currents and the coil impedance will vary with coating thickness.

Beta Backscatter Gage. The beta backscatter gage employs a radioisotope, such as promethium-147, which emits beta rays, and a detector for measuring the intensity of the beta rays backscattered by the test specimen. The intensity of the backscattered radiation is a function of the amount and kind of metals present. As the difference between the atomic number of the coating material and that of the substrate increases, the sensitivity to the weight per unit area of coating increases.

Topography

This characteristic of a coating has been discussed previously in Section 5.

Wear and Contact Resistance

Wear measurements fall into two categories, loss of metal and dimensional change. Loss of metal is easily determined by weighing the members on an analytical balance when they are small. The practical error is of the order of 0.1 mg. Radiotracer techniques are used to determine wear debris which accumulates in lubricants when the members are made radioactive. Autoradiographic techniques are useful for studying transfer.

Dimensional changes are determined by physical or optical gauging methods. Diamond stylus profile meters are commonly used. The size of the flat on the rounded end of a rider or the shrinking dimensions of an impression made with Knoop indenter in the surface of the part before sliding, are convenient ways to measure wear.

Contact resistance is the most important characteristic of switching, sliding, and joining components in electrical circuits. A modification of the 4-probe technique for measuring the conductivity of a coating is utilized. A current is impressed across the contacting members and the voltage drop is recorded and the contact resistance calculated.

Section 7

Considerations in the Selection of a Coating

The information in this section has been taken from "Metallic Coatings for Corrosion Control" by V. E. Carter..

The selection of the best coating or combination of coatings for any particular corrosion control application necessitates consideration of all the factors involved so that the most economic choice may be made consistent with the performance required. The order in which these factors should be considered is likely to be as follows:

1. The environment(s) that will be met in service

2. The service life required

3. Decorative appeal and the degree of deterioration that can be tolerated

4. The substrate material

5. The shape and size of the article

6. Any subsequent fabrication

7. Mechanical factors

However, in any particular case the order of importance may be changed to meet special circumstances.

Environmental Factors

     Consideration of the corrosive environment or environments that will be met in service generally first results in the exclusion of unsuitable coating materials, leaving a number of materials of greater or lesser merit depending on the other requirements of the application. Thus aluminum would be ruled out of consideration as a coating metal in strongly alkaline environments, aluminum and lead would be unsuitable for use in high-chloride environments, copper and zinc would be unsuitable in acidic environments—in all cases owing to their excessive rates of corrosion in these environments. Aluminum, copper, nickel and tin are resistant to atmospheric environments; aluminum and nickel are also resistant to elevated temperatures but are attacked under conditions of limited access of oxygen. Nickel, copper and tin are resistant to potable waters and sea water but aluminum is less resistant to waters, particularly when the chloride content is high. Cadmium is preferable to zinc in humid environments containing organic vapours, and aluminum, nickel and tin offer good resistance to acidic environments. Lead gives good performance in atmospheric, acidic or aqueous environments, but not in the presence of high concentrations of chlorides. Chromium remains bright and unattacked in most environments, except acidic chloride environments, but coating discontinuities may allow preferential attack on undercoats for substrate metals; on the other hand, zinc and cadmium (which are electronegative to steel in the atmosphere and in waters) can provide efficient sacrificial protection to suitable substrates— notably steel. Silver, copper and, to a lesser extent, nickel are attacked by sulphur compounds, which produce unsightly and non-protective films on their surfaces.

Service-Life Requirements and Acceptable Deterioration

The service life required affects both the choice of coating metal and also its thickness, the latter being also dependent on the severity of the environment to which it is to be exposed. It would be uneconomic, for example, to apply a coating of a highly resistant metal to a component that is required to have only a very limited service life, unless it is essential to retain an initial decorative appearance throughout that life or in cases where any risk of failure through defects could not be tolerated for any reason—e.g. safety considerations.

The ways in which choice of coating metals and their thicknesses are affected by the service environment and service life may be illustrated by reference to the accompanying tables. Thus bright nickel is unacceptable for use outdoors in exceptionally severe corrosive conditions; reduction of 15-50% in minimum nickel thickness requirement is allowed as the severity of the environment is reduced through four environmental categories; 12-16% reduction in nickel thickness is also allowed if micro-discontinuous chromium is used instead of regular chromium in outdoor service. Similarly, somewhat greater thicknesses of aluminum on steel than of zinc on steel are recommended because of the poorer sacrificial action of aluminum; thickness requirements for both types of coatings must be increased with increasing life requirements or increases in the severity of the corrosive environment.

Type and Thickness Requirements for Nickel + Chromium Coatings on Steel

Service Condition No.	Description of Typical Environment	Coating Type and Thickness	
		Ni	Cr
4	Exceptionally severe outdoor	40 μm duplex 30 μm duplex	0.3 μm regular 0.3 μm micro-discontinuous
3	Normal outdoor	40 μm bright 30 μm bright 30 μm duplex 25 μm duplex	0.3 μm regular 0.3 μm micro-discontinuous 0.3 μm regular 0.3 μm micro-discontinuous

Cont'd.

Service Condition No.	Description of Typical Environment	Coating Type and Thickenss	
		Ni	Cr
2	Indoor with condensation	20 μm bright or duplex	0.3 μm regular or micro-discontinuous
1	Indoor dry	10 μm bright or duplex	0.3 μm regular or micro-discontinuous

Effect of Substrate Material

The particular substrate material that has to be protected by a metal coating influences both the choice of coatings and also possibly the methods by which they are to be applied. Zinc and cadmium are highly effective coating metals for steel, since they are anodic to steel and provide sacrificial protection to the substrate at discontinuities in the coating. Coatings that are cathodic to a substrate metal must be applied and maintained free from defects that would expose that substrate. To ensure this the coating thickness must be sufficient to prevent corrosion penetration within the required lifetime of the component. Alternatively, cathodic coatings may be used with exposure of the substrate provided that the substrate corrosion sites will rapidly passivate by the formation of insoluble corrosion products or that the rate of attack on the substrate is insufficient to affect adversely the service life of the article. Control can also be exercised by the use of multi-layer coating systems (for example, copper or nickel undercoats with gold coatings or nickel undercoats with chromium coatings), in which case the anodic/cathodic relationship of immediate importance is that between adjacent coating layers. However, as the period of exposure of a composite system increases and corrosion penetrates through successive coating layers to the substrate, complex electrochemical relationships may be set up and one or more component of the system may be attacked at an enhanced rate.

Thickness Requirements for Zinc or Aluminum Coatings on Steel

Environment	Coating Thickness (μm) for Various Service Lives		
	5 years	15 years	50 years
Outdoor industrial	30 Zn	125 Zn	--
	50 Al	125 Al	--
Outdoor rural	7 Zn	30 Zn	--
	25 Al	37 Al	--
Outdoor marine	15 Zn	30 Zn	--
	50 Al	75 Al	--

Contd'd.

Environment	Coating Thickness (μm) for Various Service Lives		
	5 years	15 years	50 years
Indoor wet	15 Zn	30 Zn	--
	50 Al	75 Al	--
Indoor dry	7 Zn	10 Zn	15 Zn
	37 Al	50 Al	75 Al

Aluminum may be applied to steel by hot-dipping since the melting point of steel is sufficiently greater than that of aluminum, but if aluminum alloys have to be protected by pure aluminum coatings they must be applied by metal spraying or cladding. When chromium is to be applied as a coating metal, electrodeposition directly onto the substrate generally produces a coating with inadequate adhesion and/or inadequate protection of the substrate. With steel substrates nickel may be applied directly as an undercoat for chromium, but with zinc substrates an undercoat of copper must be applied beneath the nickel and with aluminum substrates it is necessary first to apply a chemical zincate treatment followed by copper and nickel undercoats. With plastic substrates it is first necessary to apply an electroless copper or nickel deposit in order to make the substrate conducting for electroplating; thick ductile undercoats are frequently necessary to ensure the retention of adhesion between the plastics substrate and the nickel and chromium layers when the effects of differential thermal expansion cause stress in the plated composites.

Effect of Shape and Size of Articles

The shape and size of the article to be coated has little, if any, effect on the choice of coating metal, except insofar as economic considerations may limit the size of article which can be coated with a given, costly material. However, shape and size influence considerably the choice of method by which the coating may be applied. Very small articles may be difficult or impossible to jig for normal electroplating; they may be coated by barrel plating techniques, by hot-dipping or perhaps by vacuum metallizing. Similarly, excessively large articles may exceed the capacity of both electroplating or hot-dipping tanks--though some latitude is possible with the latter process by using double-end dipping techniques. In these cases the only practical solution is to apply the coatings by metal spraying techniques or to redesign the article in several smaller component parts that can be coated before assembly.

Intricately shaped components (particularly those with deeply recessed regions) are difficult to electroplate with coatings of even thickness because of the limited throwing power of plating solutions--although some amelioration of this problem can be achieved, at additional cost, by using auxiliary and conforming-shape anodes to even out the current density distribution on the article being plated. Similarly, electrodeposits covering completely the inside of small-bore hollow sections can be obtained only by using internally placed anodes. Hot dipping may provide better coverage in these cases, although thickening of the coating in recessed areas may mar detailed shape and small-bore holes may become clogged with coating metal. Metal spraying techniques can cope well with irregularly shaped articles but coatings cannot be metal sprayed inside narrow bores. Chemical (electroless) plating, however, will coat the most complex shapes with even thickness both internally and externally. Probably the best way of handling both complex-shaped and excessively large components in order to achieve the best coatings is to redesign them so as to simplify the application of the chosen coating by the desired method. Indeed it is a fundamental principle in achieving both the best application and the best performance of coated metal components that the coating requirements be taken into account fully at the original design stage.

Effect of Subsequent Fabrication

Fabrication that must take place after application of any metal coating should always be considered when both the coating metal and its method of application are chosen. Obviously any cutting or trimming that has to be done after applying the coating will damage the coating and expose the substrate metal. Anodic coatings may well be able to cope with the exposed area of substrate by providing sacrificial protection, provided that the area in question is not too large, but the increased rate of consumption of the coating metal that results from the presence of exposed substrate may well markedly reduce the ultimate effective life of the coated article compared with that which would have been achieved if exposure had been avoided. In the case of cathodic coatings, however, any substrate metal exposed as a result of cutting after coating will itself be preferentially attacked; provision therefore has to be made to provide local protection in the exposed area or to repair the coating before the article is placed in service. The only coating process that can be readily applied *in situ* to a limited area of a large structure, thus repairing any damaged coatings, is metal spraying (although it may be possible with certain coating metals such as tin, lead and their alloys to effect localized repairs by soldering or brazing techniques).

Post-coating welding operations destroy the coating in the weld zone, and in part or perhaps the whole of the heat-affected zone, so local repair or protection is required similar to that necessary where cutting has taken place. In addition, the coating metal may well affect the welding process by alloying, causing unsound or embrittled weldments. A further hazard during welding can arise from the production of toxic vapors produced from the coating metals; for this reason cadmium should never be chosen as a coating metal for steel that must be subsequently welded.

Assembly of coated components may produce creviced regions, such as in bolted-up lap joints or beneath the heads of fasteners, and the susceptibility of the chosen coating metal to crevice corrosion must be borne in mind. Similarly, bi-metallic contact can occur on assembly; ideally, such contacts should be designed out of the structure, or assembly made with non-metallic (insulating) separators in the joint, but where these methods cannot be followed the coating metal chosen must be as compatible as possible with the dissimilar contacting metal. For example, where steel and aluminum must be in contact the steel should be coated with cadmium, since cadmium and aluminum when in contact do not lead to bi-metallic corrosion of the latter metal.

Mating and threaded components must be designed and produced so as to allow for dimensional changes occurring during coating, and the coating thickness and method of application must be chosen to achieve the best compromise between adequate fit and adequate protection with the minimum of post-coating machining.

The internal stress, ductility and brittleness of coating metals (and, where appropriate, of alloy layers) must be taken into account when choosing a coating metal and its method of application for a component that must be deformed during fabrication or in service. Electrodeposits such as chromium and some nickels can withstand only a small amount of deformation without cracking or spalling; the development of excessively thick alloy layers during hot-dipping also embrittles the coating and leads to failure on deformation. The hardness, ductility and frictional properties of a coating metal may be of considerable consequence in post-fabrication. A very soft coating such as lead, or to lesser extent aluminum, can deform readily under load; this may lead to more efficient elimination of some crevices but may also cause localized thinning of the coating or even exposure of the substrate. Sprayed zinc or aluminum coatings on steel are of special value in applications where friction grip bolting is involved. Slip factors of the order of 0.45-0.55 are readily obtained with sprayed zinc coatings, and in the case of sprayed aluminum coatings the slip factor can rise as high as 0.7. Galvanized steel in the "as galvanized" condition has a somewhat lower slip factor than sprayed zinc, owing to the smoothness of the deposit, but in service under dynamic loading a hysteresis loading cycle occurs which produces self-roughening of the faying surfaces with a consequent locking action which prevents any slip taking place. Conversely, the low torque resistance of cadmium makes it the best choice of coating metal for steel bolts for structures that must be assembled and dismantled frequently.

One point not directly concerned with fabrication but allied to it—and a matter that is frequently overlooked—is to ensure that all components in a composite structure have comparable effective service lives. Thus hot-dipped galvanized components having a coating thickness of some 50 μm or more may be assembled using fasteners that have been electroplated with zinc to a thickness of perhaps only 10-20 μm. In cases such as this the life of the fasteners will be only 20-40% that of the rest of the structure (since the life of a zinc coating is proportional to its thickness) and unsightly rusting, or perhaps even collapse, will occur permaturely.

Mechanical Factors

Mechanical factors that must be considered when choosing a coating are mostly those of stress during service—either dynamic or static. The application of heat during hot-dipping processes, and to a lesser extent during metal spraying, can adversely affect the mechanical properties of the substrate metal by partial or complete annealing during coating. If this occurs, the strength of the completed component may be inadequate for its application, or the component may be distorted during coating so that subsequent assembly is difficult or even impossible.

When coated components are stressed during assembly or in service, failure can occur if the coating metal is susceptible to stress corrosion—

for example, stressed copper or copper alloys exposed to ammoniacal environ-ments. Alternatively, a substrate metal susceptible to stress corrosion may be completely protected by means of a suitable metal coating—for example, high-strength aluminum alloys coated with sprayed pure aluminum or with zinc. Dy-namic stressing during service may produce flexing of a component, and in these cases if the coating is brittle it may crack and expose the substrate with consequent loss of protection; an example of this may be seen in the case of thick 'crack-free' chromium deposits, which fracture through brittleness when flexed (as in motor car bumper-bars or hub-discs), the cracks then propa-gating through the nickel undercoat to expose the steel substrate.

Coating by electrodeposition, by the nature of the process, frequently produces cathodic hydrogen at the metal surface, and this hydrogen may be absorbed in the coating and/or in the substrate. The presence of this hydro-gen in certain metals can result in embrittlement, such as is the case with high-tensile steels, leading to brittle fracture when stressing occurs in service. Provisions should be made in the appropriate standard specifications to carry out stress-relief annealing treatments to remove or minimize these effects. Thus BS 1224:1970 (Electroplated coatings of nickel and chromium) specifies stress-relief annealing before plating for 1 hour at 130-210°C ac-cording to the type of steel, or after plating for 5 hours at 190-210°C or for 15 hours at 170°C if the higher treatment temperature would be harmful to the mechanical properties of the steel. Similar provisions are made for zinc or cadmium plated high-strength steel components, but in special cases it might be desirable to avoid coating by electrodeposition and use either metal sprayed zinc or vacuum metallized zinc or cadmium coating processes, thus avoiding exposure to hydrogen.

Sprayed metal coatings may also be preferable in applications where fatigue loading is involved, since the compressive stressing of the surface layers of the substrate by the grit-blasting pretreatment may improve fatigue properties. In applications where fretting of bolted joints can occur, the rough irregularities present on the surface of a sprayed metal coating can increase friction in the joint and so reduce fretting corrosion.

Metal-coated plastics materials are subjected to particular hazards through mechanical forces during service. The principal of these is the risk of rupture between the coating and the substrate through stresses arising during temperature changes because of the great disparity between the coefficients of thermal expansion of metals and plastics. In practice it may be necessary to incorporate a sufficient thickness of a ductile undercoat such as copper, which will prevent rupture when differential expansion and contraction occur. Cases have also occurred in practice where plastics components plated with nickel and chromium have failed under stress in service by fracture in a sharply angled recess, although similar plastics components used unplated have performed satisfactorily. This mode of failure is caused by stress concentration in the notch of the recess cracking the chromium coating, the crack then propagating through the metal undercoats and through the plastics substrate. In such cases the only remedy lies in the redesign of the component to eliminate the notch effect.

Mechanical factors other than stress that must be considered are those involving movement. Relative movement between coated components and other components in an assembly occasions wear, and the wear resistance of the coating metal is then the primary consideration. In general the hardness of the metal coating is the principal factor of concern here, since normally a softer material wears when in moving contact with a harder material (although there are exceptions to this). Diffusion coatings and electrodeposited coatings of hard metals such as chromium and nickel are principally used for wear-resistant applications, but metal sprayed coatings subsequently machined and perhaps also heat-treated are often employed. A secondary property of metal coatings of importance for wear resistance is the adequacy of its adhesion to the substrate. Where adhesion is poor, rubbing action can cause localized rupture at the coating/substrate metal interface, leading to blistering or even complete spalling off of the coating metal.

Effect of Environmental Movement

Movement of the environment may also need to be considered. In applications involving exposure to moving liquids or gases, erosion can occur. This may be insufficient to erode the metal itself but nevertheless sufficient to remove the protective films locally and thus set up anodic areas where enhanced corrosion will occur (for example, impingement corrosion of copper and its alloys immersed in moving water), or it may be of sufficient magnitude to damage mechanically the metal itself (as in cavitation corrosion). In either case, premature localized penetration through a coating may occur, causing exposure of the substrate with consequent loss of protection, or even stripping the coating completely from considerable areas of the component as corrosion undercuts the coating and turbulence is increased in the moving liquid. The remedy in these cases lies in the selection of a coating material (e.g. nickel or nickel alloy) that offers improved resistance to erosion, or in the redesign of the component so that the erosive effects are reduced.

Section 8

Corrosion Principles

A typical corrosion process, the reaction of iron with HCl under anaerobic conditions, is represented by the equation:

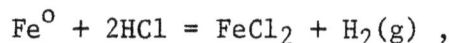

$$Fe^o + 2HCl = FeCl_2 + H_2(g) \ ,$$

although a more accurate representation is:

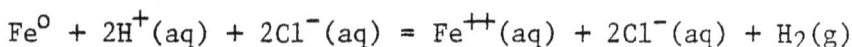

$$Fe^o + 2H^+(aq) + 2Cl^-(aq) = Fe^{++}(aq) + 2Cl^-(aq) + H_2(g)$$

which indicates that the acid reactant and the $FeCl_2$ corrosion product exist largely in the ionic state. The reaction as written requires that an iron atom in the surface and two hydrated protons in solution, along with two hydrated chloride ions in the immediate vicinity, have the proper energetic and geometrical arrangement to react to form a hydrated ferrous ion and a hydrogen gas molecule. The probability of this reaction to occur by the mechanism implied by the equation at room temperature is very low because too many species must meet simultaneously. We may, however, conceive of this reaction occurring in two independent steps.

Anodic Reaction: $Fe - 2e^- = Fe^{++}(aq)$

Cathodic Reaction: $2H^+(aq) + 2e^- = H_2(g)$

Thus, electron loss occurs at one site and electron gain occurs at another. The overall reaction takes place at a rate determined by the more difficult of these two steps. This simple concept— independent anodic and cathodic half-reactions— is the basis of the electrochemical mechanism of corrosion. It is the purpose of this chapter to develop this concept and to show its utility in understanding many corrosion processes. The following chapters will use these concepts to interpret the real-life behavior of metals under corrosive conditions.

When the anodic and cathodic reactions are physically separated and an electrical conductor connects the two separated reactions, electrons will flow in the external circuit and electrical work can be performed. The ordinary flashlight cell represents a useful application of this principle as shown in Figure 8-1. The outside casing of the cell is made of zinc which serves as the anode and the central rod of graphite surrounded by finely divided carbon and manganese dioxide serves as the cathode. The space between the anode and cathode is filled with a porous non-conductor moistened with a

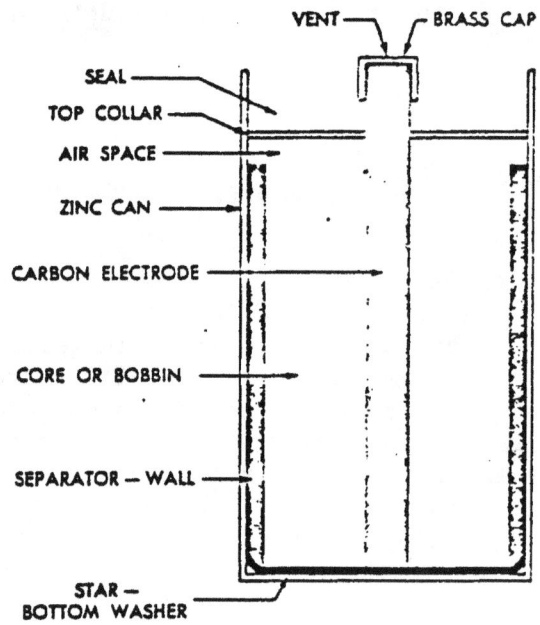

VENT — — BRASS CAP

SEAL —

TOP COLLAR —

AIR SPACE —

ZINC CAN —

CARBON ELECTRODE —

CORE OR BOBBIN —

SEPARATOR — WALL —

STAR —
BOTTOM WASHER —

Figure 8-1. Schematic representation of the
ordinary flashlight cell.

solution of ammonium chloride. Electrons are supplied to the external circuit by the reaction of the zinc anode:

$$Zn - 2e^- = Zn^{++}$$

The ammonia supplied by the NH_4Cl electrolyte keeps the zinc ion in a solubilized form as a zinc-ammonia complex. The electrons are consumed at the cathode by the reaction:

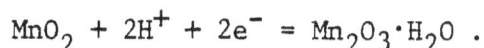

$$MnO_2 + 2H^+ + 2e^- = Mn_2O_3 \cdot H_2O \ .$$

The concentration of the ions in the electrolyte and the overall chemical reaction provide a cell driving force of the order of 1.5 V when the current drawn is low. An important requirement of the flashlight cell is that there be very little local cell reaction on the zinc electrode in the absence of external contact between the zinc and graphite. The success of this cell resides in the art of the manufacturer in devising means to reduce the corrosion of zinc when current is not being drawn from the cell.

Much of the science of corrosion is based on understanding how and why the anodic and cathodic reactions are separated on a metal surface exposed to an electrolyte. That they often are separated is shown dramatically by the experiment schematically represented in Figure 8-2. An iron nail is immersed in a solution containing sodium chloride and dissolved oxygen, to which is added a low concentration of potassium ferricyanide and phenolpthalein indicator. Convective flow of the liquid is minimized by making a gel. As shown in the figure, the points and the heads of the nail are surrounded by a dark blue zone of ferrous ferricyanide and the central part of the nail is surrounded by a pink color indicative of an alkaline condition. Thus, the anodic reaction, $Fe - 2e^- = Fe^{++}$, takes place at the point and head of the nail and the cathodic reaction, $H_2O + 1/2O_2 + 2e^- = 2OH^-$, takes place on the central part of the nail. The corrosion process in this case may be visualized as an electrochemical cell, but the current is not supplied to an external circuit but rather flows between two adjacent regions of the same piece of metal.

When a piece of iron is immersed in oxygen-free 0.1N acid at room temperature and the potential of the iron is measured against a reference electrode, as schematically shown in Figure 8-3, a value of the order of -0.3 V (SHE) is obtained. The potential is known as the corrosion potential and represents the electrochemical state of the metal in that particular environment under dynamic (non-equilibrium) conditions. How is this potential to be visualized in terms of independent anodic and cathodic reactions? Since the solution is oxygen free, the reactions we will be concerned with are:

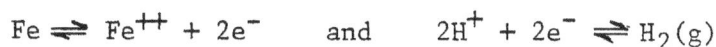

$$Fe \rightleftharpoons Fe^{++} + 2e^- \quad \text{and} \quad 2H^+ + 2e^- \rightleftharpoons H_2(g)$$

Once a small amount of reaction has occurred we may consider that the iron/solution interface contains elemental iron, ferrous ions, hydrogen gas and

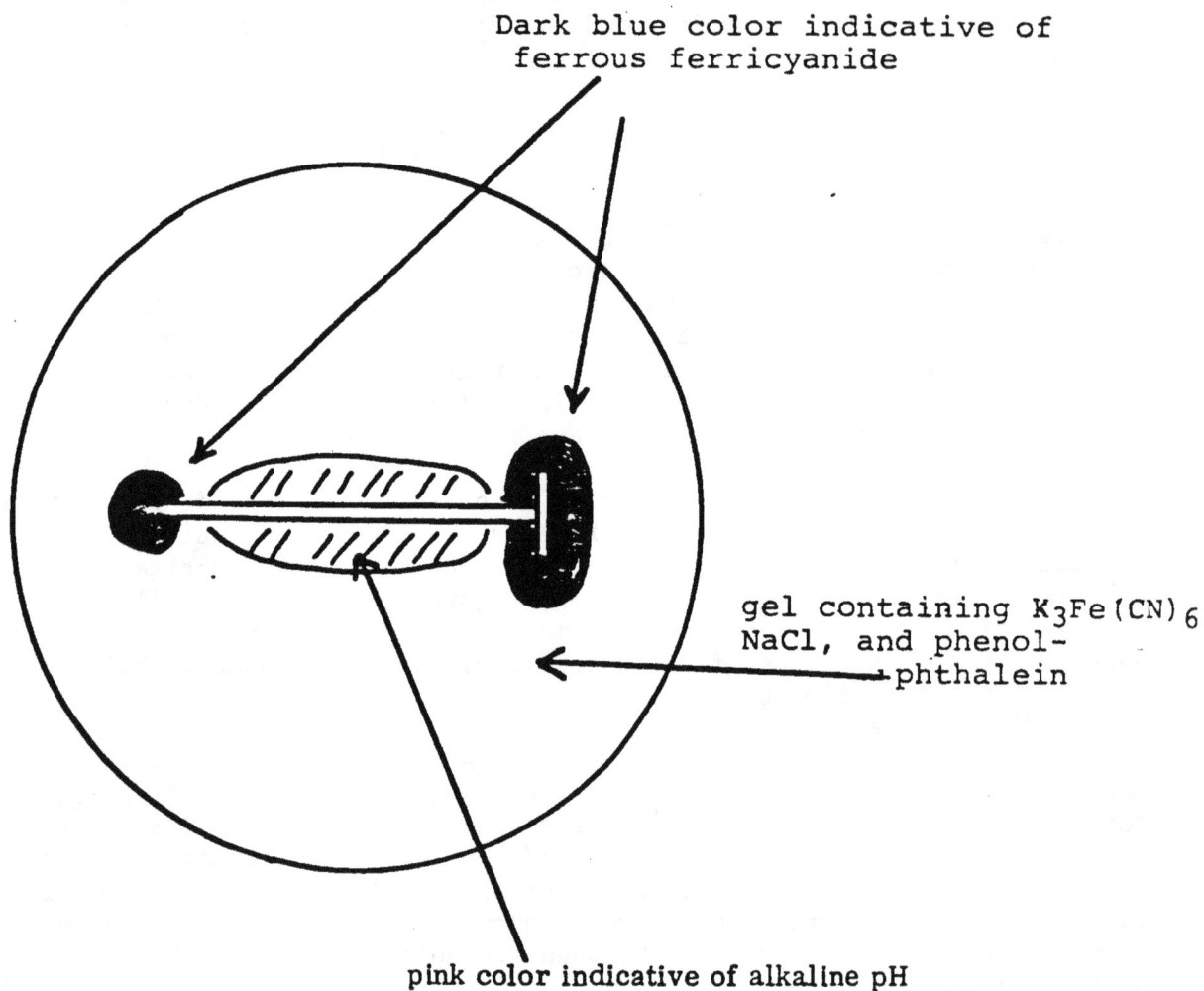

Figure 8-2. Corrosion of an iron nail in a gel containing NaCl and $K_3Fe(CN)_6$ and phenolphthalein indicators. Anodic reaction occurs at point and head of nail and cathodic reaction occurs in central part of nail.

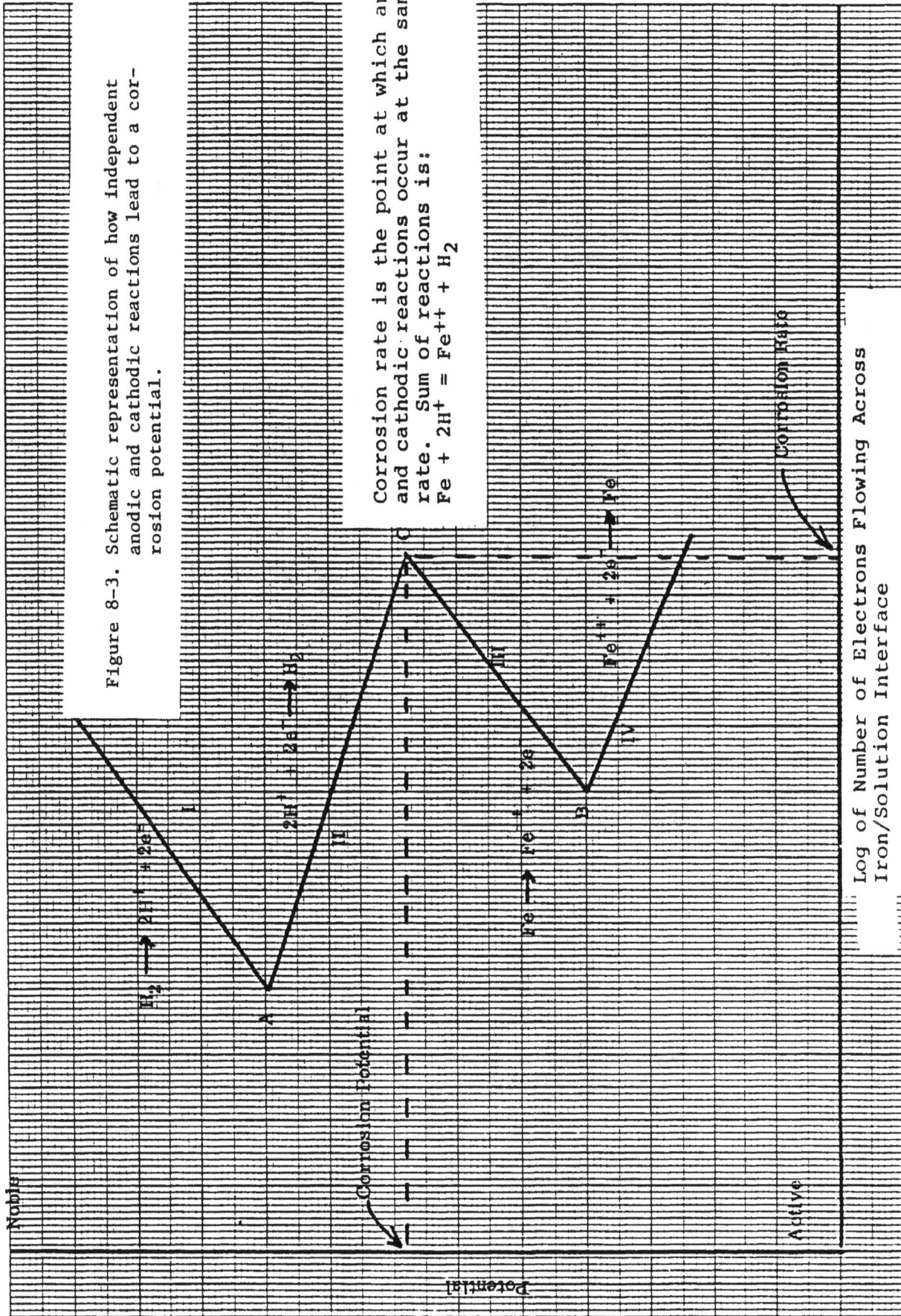

Figure 8-3. Schematic representation of how independent anodic and cathodic reactions lead to a corrosion potential.

Corrosion rate is the point at which anodic and cathodic reactions occur at the same rate. Sum of reactions is:
$$Fe + 2H^+ = Fe^{++} + H_2$$

$H_2 \rightarrow 2H^+ + 2e^=$

$2H^+ + 2e^- \rightarrow H_2$

$Fe \rightarrow Fe^{++} + 2e^=$

$Fe^{++} + 2e \rightarrow Fe$

Corrosion Rate

Corrosion Potential

Potential

Active

Log of Number of Electrons Flowing Across
Iron/Solution Interface

hydrogen ions, plus an equivalent of anions. We may also consider that the reaction tendency is represented by a specific potential, denoted A for the hydrogen-hydrogen ion equilibrium and denoted B for the elemental iron-ferrous ion equilibrium. Under these hypothetical steady-state conditions, there is a flow of electrons across the iron/solution interface known as the exchange current density. Now we may further imagine that an internal force, represented by a potential, is applied to drive the reactions from the steady-state situation. The departure from points A and B as a greater number of electrons cross the iron/solution interface is represented schematically by lines I and II for the hydrogen-hydrogen ion system and by lines III and IV for the elemental iron-ferrous ion system. The intersection of lines II and III at point C is where the electron-accepting and electron-donating reactions occur at the same rate. This is the corrosion rate.

It should be recognized that the only experimental information that was used in the construction of Figure 8-3 was the measured corrosion potential and the corrosion rate as determined by an analytical method. The constructions to the left of point C are schematic in nature and are based on the assumptions inherent in the electrochemical mechanism of corrosion. Another experimental approach is available, however, that is very useful in understanding the fundamentals of corrosion science. An external potential may be applied to the iron electrode such that it is made more anodic in one case and more cathodic in a second case. If the potential of the iron electrode is determined as a function of the current flowing across the interface according to the scheme shown in Figure 8-3, values are obtained so that a plot of potential vs. current may be constructed. Data obtained by Sayano and Nobe for iron in 1N H_2SO_4 in the absence of oxygen are summarized in Figure 8-4. It will be noted that beyond a certain current density there is a linear relationship between the applied potential and the logarithm of the current density. The relationship, $\phi = \beta \log i$, has been observed in many systems and the value of β is termed the Tafel slope after an electrochemist who studied the potential current density relationship many years ago. The values of β_A, the slope of the anodic curve, and β_C, the slope of the cathodic curve, vary with the metal being studied and with the electrolyte in which the polarization curves are determined. The slopes are sensitive to small amounts of impurities on the metal surface, reducing or oxidizing materials in the solution, and inert contaminants, so it is not surprising that the Tafel slopes in similar systems determined in different laboratories are often at variance with one another.

Many situations exist in which two metals are in intimate contact with each other and with the electrolyte. Under these conditions it is likely that one metal will be more active for the cathodic hydrogen evolution reaction. The electrochemical concept of corrosion allows us to understand the consequences of a bi-metallic couple. Figure 8-5 shows what happens to the corrosion rate of tin in boiling 2N HCl under anaerobic conditions when the tin is contacted with another metal, either a solid of equal surface area or by adding cations to the solution that react with the tin to form elemental metal. In the absence of a second metal the corrosion of tin under these conditions is of the order of 2 mg/cm^2/hr and, when tin is contacted with an equal surface area or iridium metal, the corrosion rate is increased to approximately 100 mg/cm^2/hr. When iridium metal is deposited on the surface by a metal replacement reaction, the rate is increased to approximately 1000 mg/cm^2/hr. Figure 8-6 explains in a schematic way why this occurs. The exchange current density for the hydrogen ion/hydrogen reaction on tin is very low and the

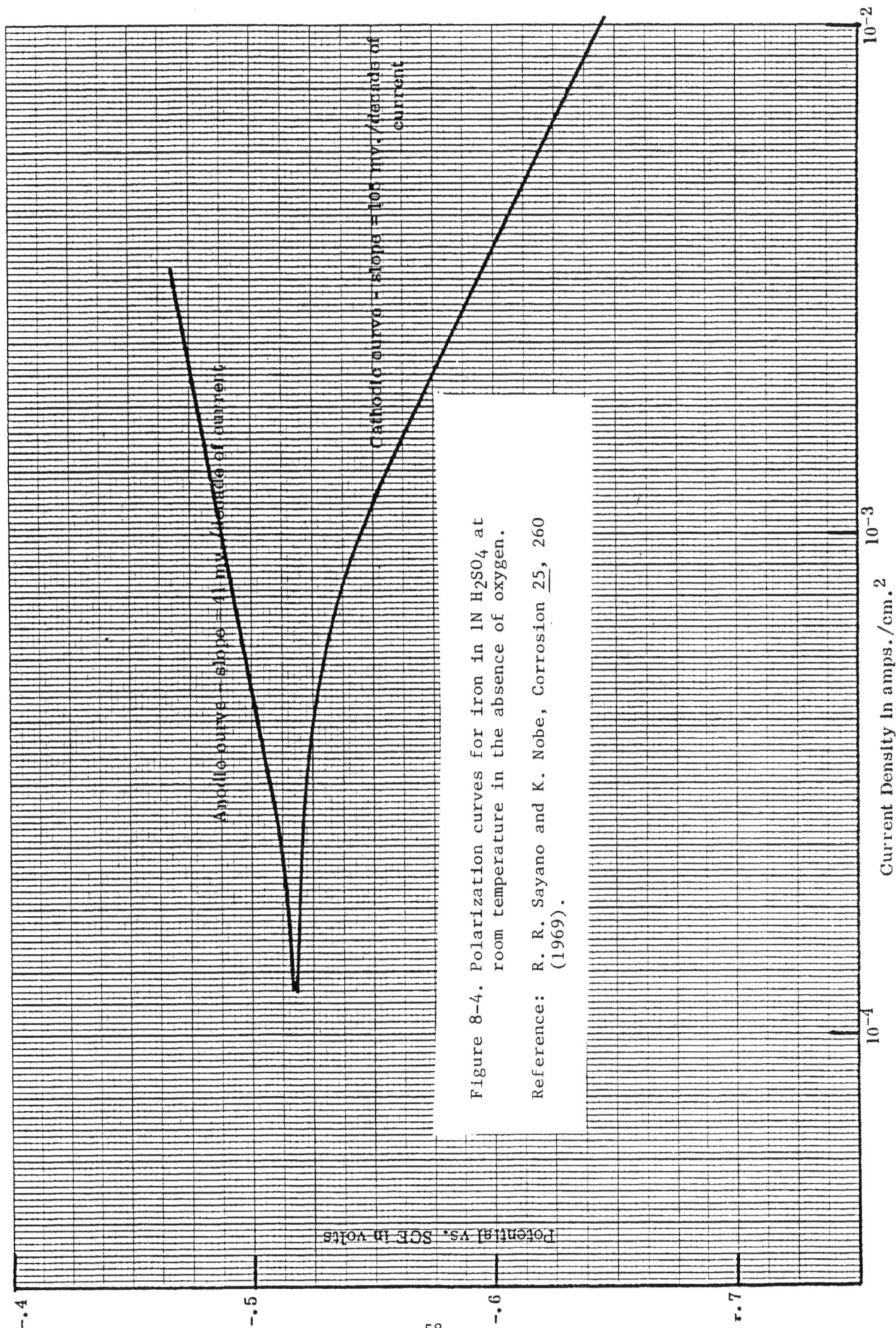

Figure 8-4. Polarization curves for iron in 1N H_2SO_4 at room temperature in the absence of oxygen.

Reference: R. R. Sayano and K. Nobe, Corrosion 25, 260 (1969).

Current Density in amps./cm.^{2}

Potential vs. SCE in volts

Cathodic curve - slope = 105 mv./decade of current

Anodic curve - slope = 41 mv./decade of current

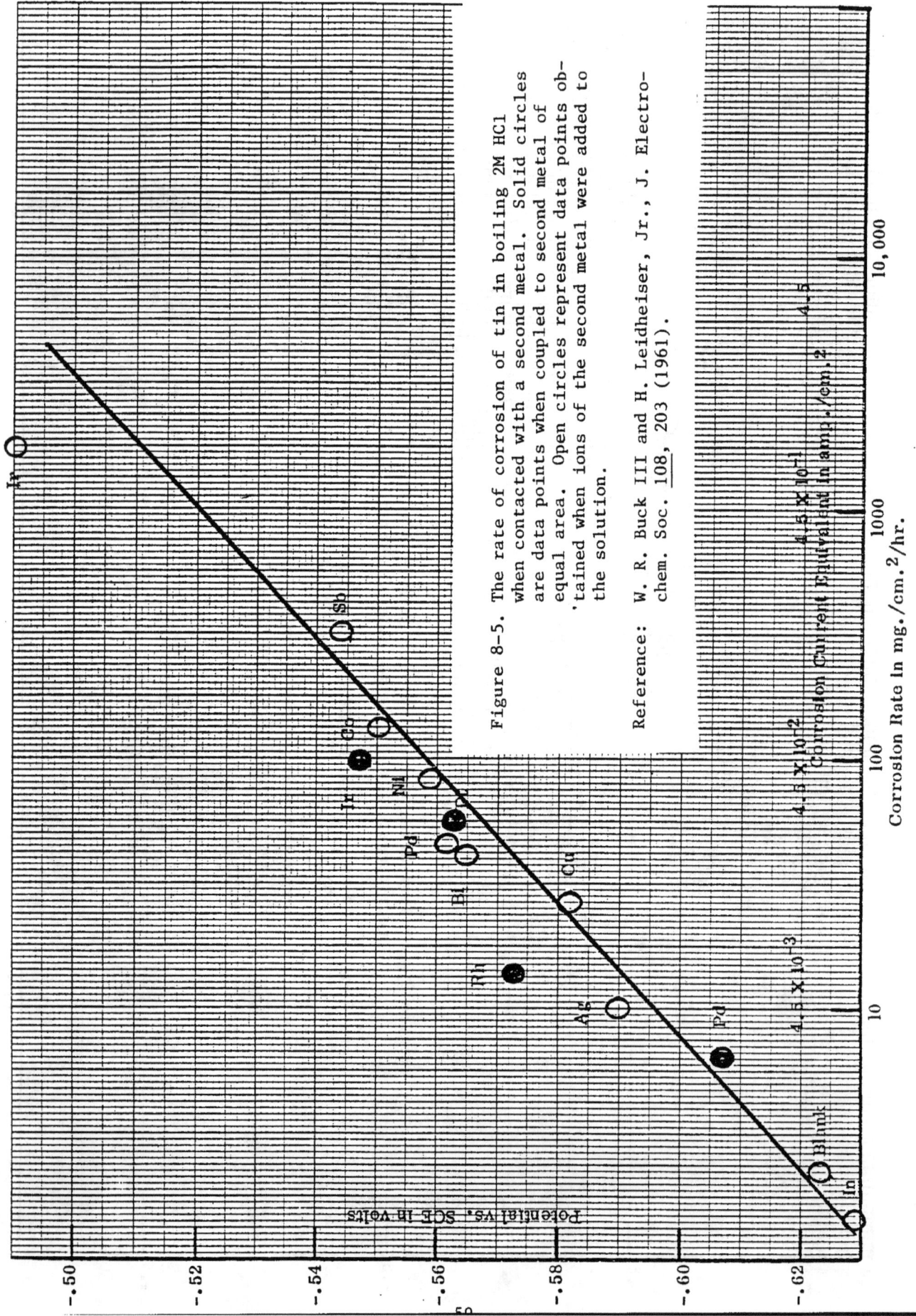

Figure 8–5. The rate of corrosion of tin in boiling 2M HCl when contacted with a second metal. Solid circles are data points when coupled to second metal of equal area. Open circles represent data points obtained when ions of the second metal were added to the solution.

Reference: W. R. Buck III and H. Leidheiser, Jr., J. Electrochem. Soc. 108, 203 (1961).

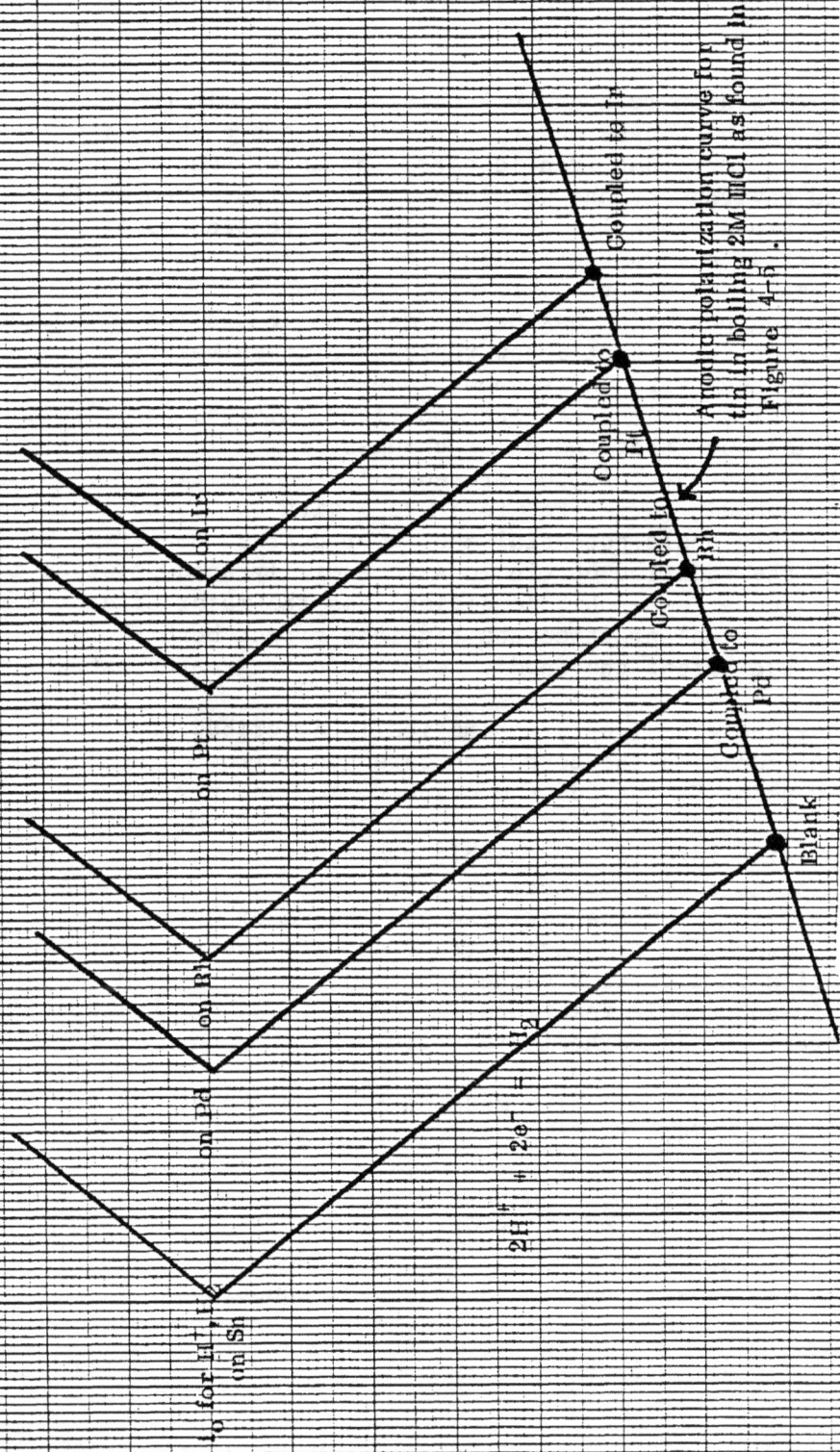

Figure 8-6. Schematic explanation of the experimental results summarized in Figure 8-5. Note: The low current density for the exchange current of the hydrogen ion/hydrogen reaction on palladium is caused by the poisoning of the palladium surface by the tin ions in solution.

Exchange Current Density

Potential

intersection of the cathodic polarization curve with the anodic polarization curve is at the point marked "blank". When tin is coupled to palladium, rhodium, platinum or iridium of equal surface area, the corrosion rate is greatly increased because the cathodic curve intersects the anodic curve at a higher current density.

Measurements of corrosion rate and potential often permit conjectures about the mechanism by which changes occur. Two examples are shown in Figures 8-7 and 8-8. Consider that iron is corroding under anaerobic conditions in a mild acid environment and we measure the corrosion potential and the corrosion rate experimentally. These values correspond to point C in Figure 8-7. Now let us add a very small amount of a cation, such as Sn^{++}, to the solution, and again measure the corrosion rate and corrosion potential. We find that the corrosion potential has become more noble and the corrosion rate has been reduced such that point F now represents these values. The rationalization of these observations is shown schematically in the figure. The exchange current density for the ferrous ion/elemental iron is affected most by the Sn^{++} addition such that the value changes from point B in the absence of the additive to point D in the presence of the additive. Thus, the intersection of the anodic curve ($Fe \rightarrow Fe^{++} + 2e^-$) with the cathodic curve ($2H^+ + 2e^- \rightarrow H_2$) is at a more noble potential and lower current density. We may thus conclude that the addition of tin ions to a mildly acid solution in which iron is corroding under anaerobic conditions has the major effect on the anodic reaction. In actual practice the tin ions affect both the anodic and cathodic reaction, but affect the anodic reaction to a greater extent.

Consider now a second experiment, schematically described in Figure 8-8. In the absence of additive the corrosion rate and corrosion potential are indicated again by point C. We add the organic compound pyridine to the solution such that its concentration is approximately $10^{-5}M$. It is observed that the corrosion rate is appreciably reduced and the corrosion potential is changed in the active direction. These observations can be accounted for by recognizing that the pyridine has the major effect on the cathodic reaction and lowers the exchange current density for the hydrogen ion/hydrogen gas reaction on iron to a value denoted by point D. The intersection of the cathodic and anodic reactions is now at point F. Pyridine thus has the major effect on the cathodic reaction.

The foregoing remarks have all been made on the basis that the cathodic reaction is $2H^+ + 2e^- = H_2$. This reaction is the dominant one in very acidic solutions, in confined places to which oxygen has little or no access, and in environments in which oxygen is purposely excluded. However, most corrosion processes occur in the atmosphere where oxygen is present and other cathodic processes occur with greater facility. Some of these alternate cathodic reactions may be illustrated by the following equations:

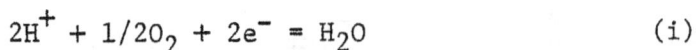

$$2H^+ + 1/2O_2 + 2e^- = H_2O \qquad\qquad (i)$$

Reaction (i) is particularly important with metals such as tin since the hydrogen evolution reaction on tin has a large activation energy barrier.

Figure 8-7. Schematic diagram of the effect of an
additive that inhibits the anodic reaction selectively.

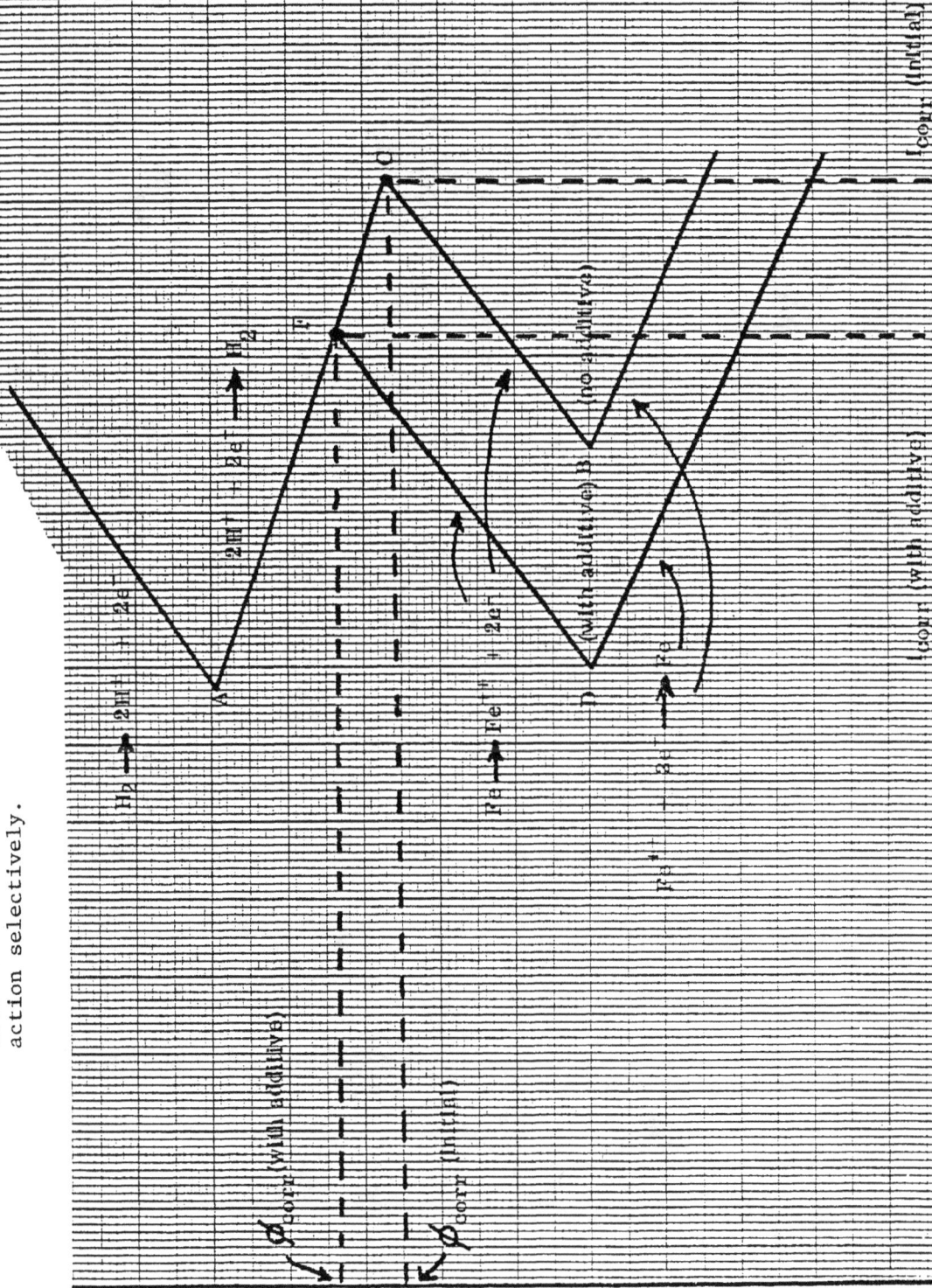

Log of Current Density

Potential

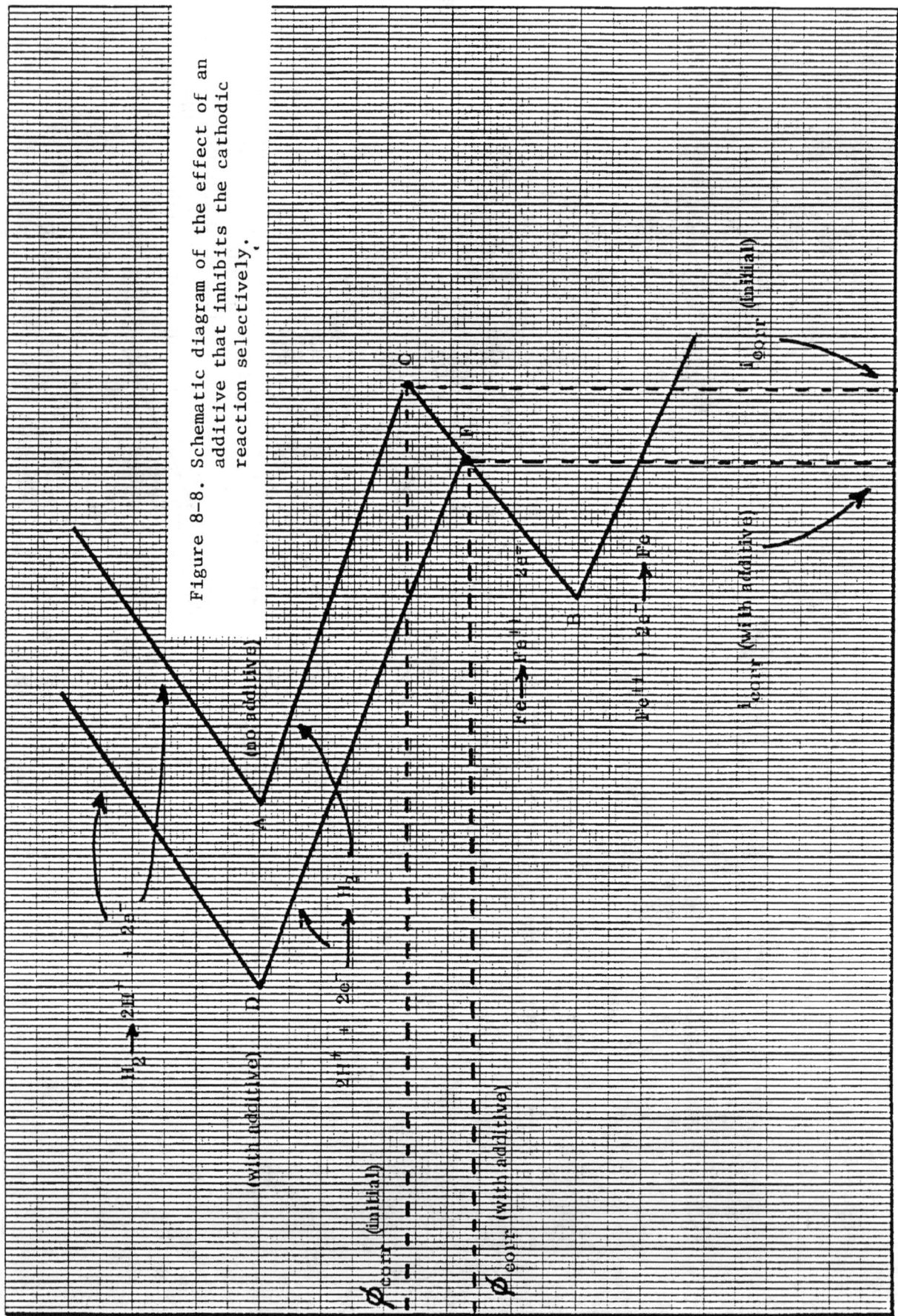

Figure 8–8. Schematic diagram of the effect of an additive that inhibits the cathodic reaction selectively.

Log of Current Density

Potential

$$2H^+ + O_2 + 2e^- = H_2O_2 \qquad \text{(ii)}$$

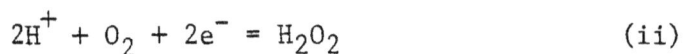

Reaction (ii) is an alternate to reaction (iii) when the pH of the medium becomes high.

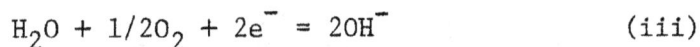

$$H_2O + 1/2O_2 + 2e^- = 2OH^- \qquad \text{(iii)}$$

Reaction (iii) is the common cathodic reaction that occurs when metals corrode in aqueous environments exposed to the air. It is also the common reaction that occurs beneath coatings on metals when the coating has a finite permeability to oxygen.

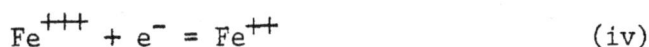

$$Fe^{+++} + e^- = Fe^{++} \qquad \text{(iv)}$$

and

$$Cu^{++} + e^- = Cu^+ \qquad \text{(v)}$$

Cathodic reactions involving metal ions may occur under certain circumstances. These reactions tend to become important when portions of the electrolyte in contact with the metal surface are exposed to oxygen and other portions of the metal surface are exposed to electrolyte in which the oxygen concentration is low. Reactions involving the chromate ion are also important in some phases of corrosion inhibition, as for example:

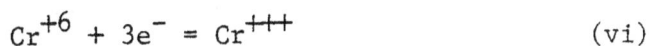

$$Cr^{+6} + 3e^- = Cr^{+++} \qquad \text{(vi)}$$

Passivity

The phenomenon of passivity has captured the attention of many famous electrochemists. Passivity was apparently discovered by Keir in 1790 and both Schönbein and Faraday were fascinated by passivity in the early 1800's. A general discussion of passivity was organized by the Faraday Society in 1913 [8-1] and the interest continues as evidenced by the fact that the 5th International Conference on Passivity was held in Bombannes, France in June 1983, and the proceedings were published as a book.

When an active metal such as iron is polarized anodically in environments in which the ion is soluble, the polarization curve is often represented by a straight line on a potential vs. log current plot. In some environments, however, the linear portion of the plot is limited to a certain potential range and the total polarization curve has a shape such as shown in Figure 8-9. This curve is obtained under steady-state conditions when sufficient time is allowed between each measured point. If the potential is continuously changed at a fixed rate, a curve of the type shown in Figure 8-10 may be obtained. The corrosion potential in Figure 8-9 is at point A and the linear range for the potential vs. log current plot extends from B to C. At point D the current passing through the interface at the anode surface drops to a very low value at point E. Between E and F, a significant change in potential has little effect on the current. Common terms used to describe the curve are:

(a) The region between B and C is known as the active range.

(b) Point D is known as the critical current and critical potential for passivity or more simply, the passivation potential.

(c) The region between E and F is known as the passive range.

(d) The region beyond F is known as the transpassive region. The sharp rise in current at F can result both from oxygen evolution and from metal dissolution.

When a metal yields a polarization curve of the same general shape as the ideal curve represented in Figure 8-9, the metal is said to exhibit passive behavior. Measured polarization curves for iron, cobalt and nickel in a borate buffer are given in Figures 8-11 - 8-13. Representative curves for other systems are given at the end of the section. The exact shape of the measured curves is a function of the method of measurement. The values for the current flowing in the passive range are a function of the rate at which the potential is changed as well as the potential at which the measurement is made. Representative values for the passive current in several systems are listed in Table 8-1.

Much of the research on the passivity of iron has been carried out either in 1N H_2SO_4 or in a boric acid-borate buffer at pH of approximately 8.4. For simplicity the discussion below will concentrate on studies carried out in the borate buffer. The polarization curve as determined by

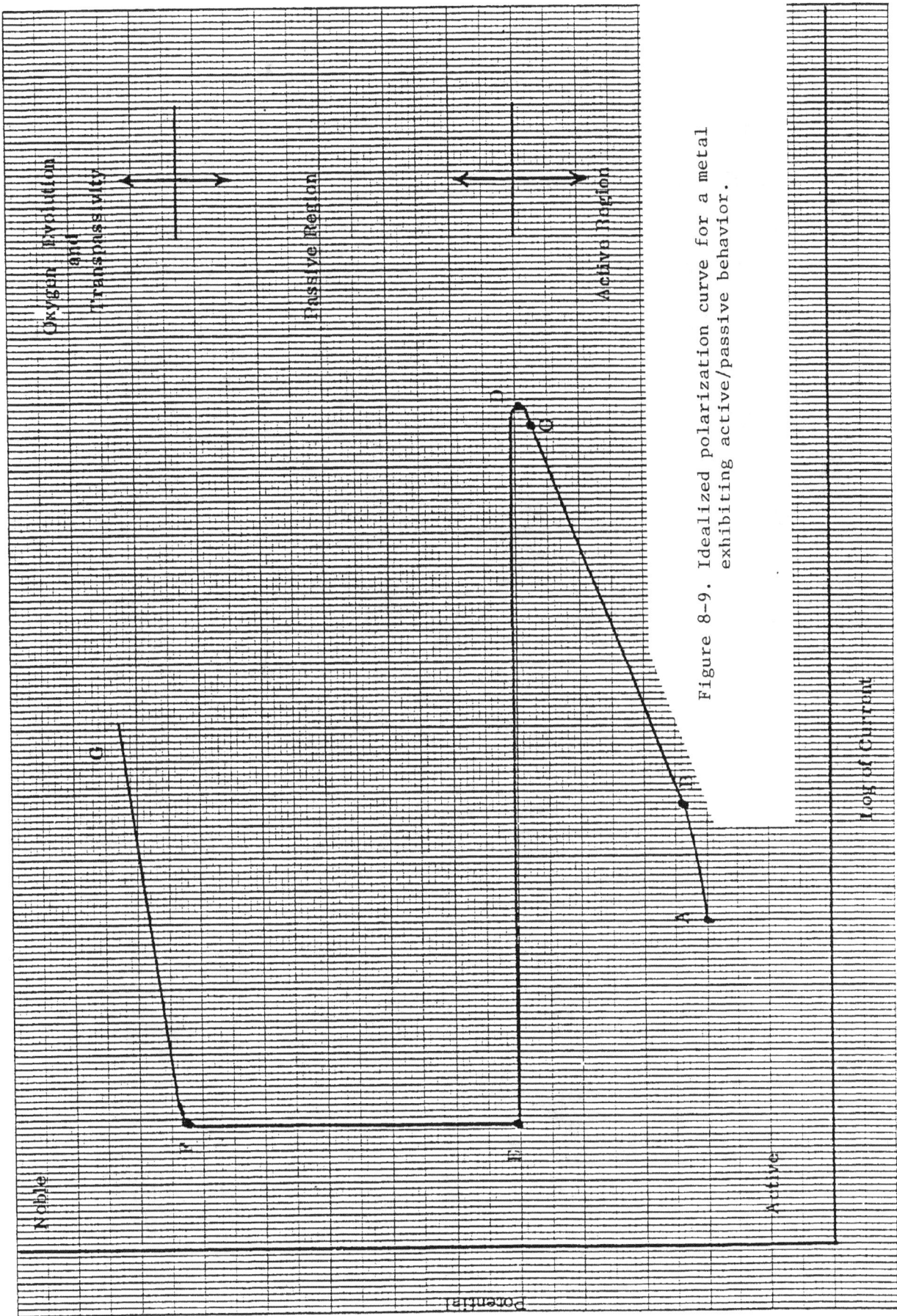

Figure 8-9. Idealized polarization curve for a metal exhibiting active/passive behavior.

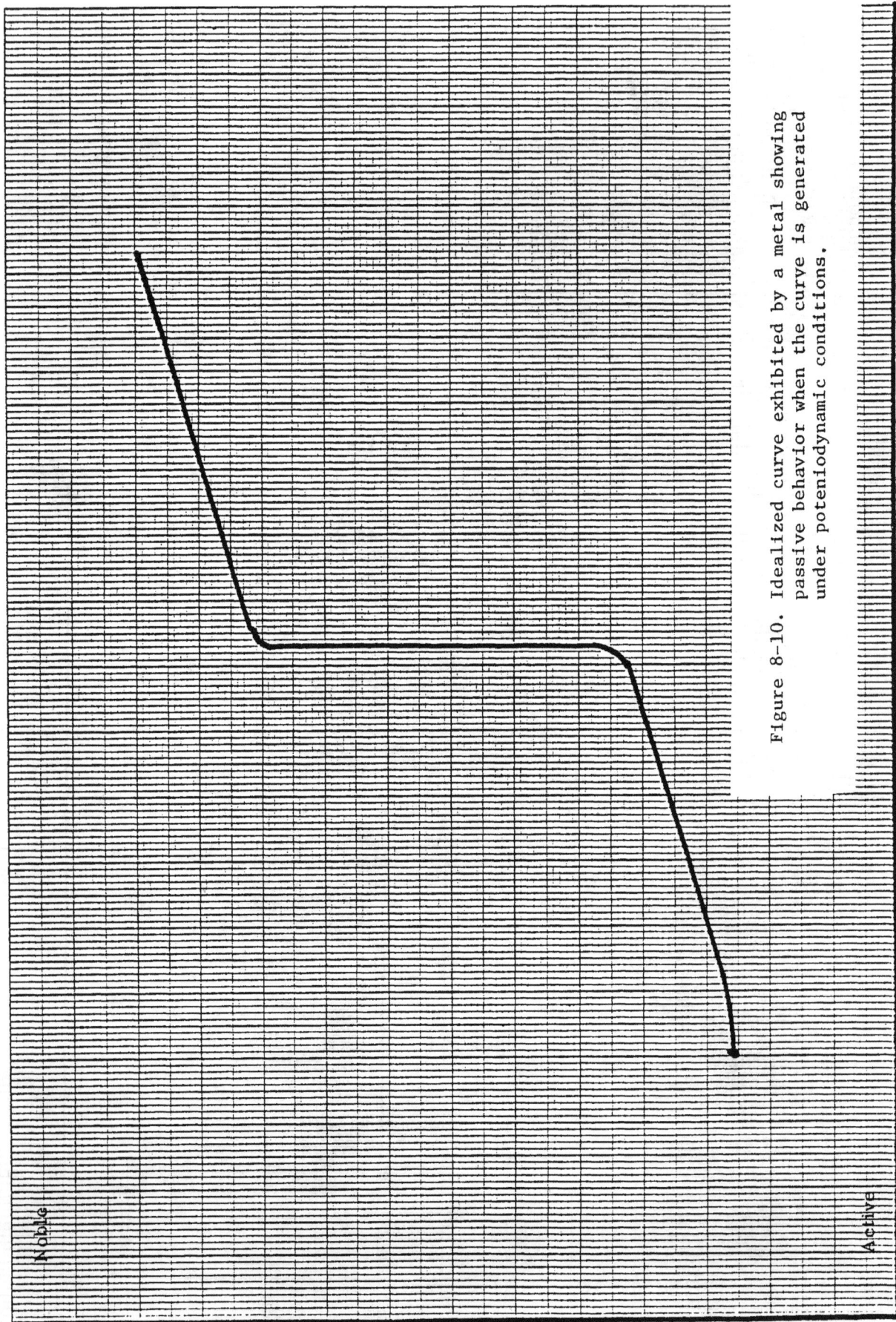

Figure 8-10. Idealized curve exhibited by a metal showing passive behavior when the curve is generated under potentiodynamic conditions.

Noble

Active

Potential

Log of Current

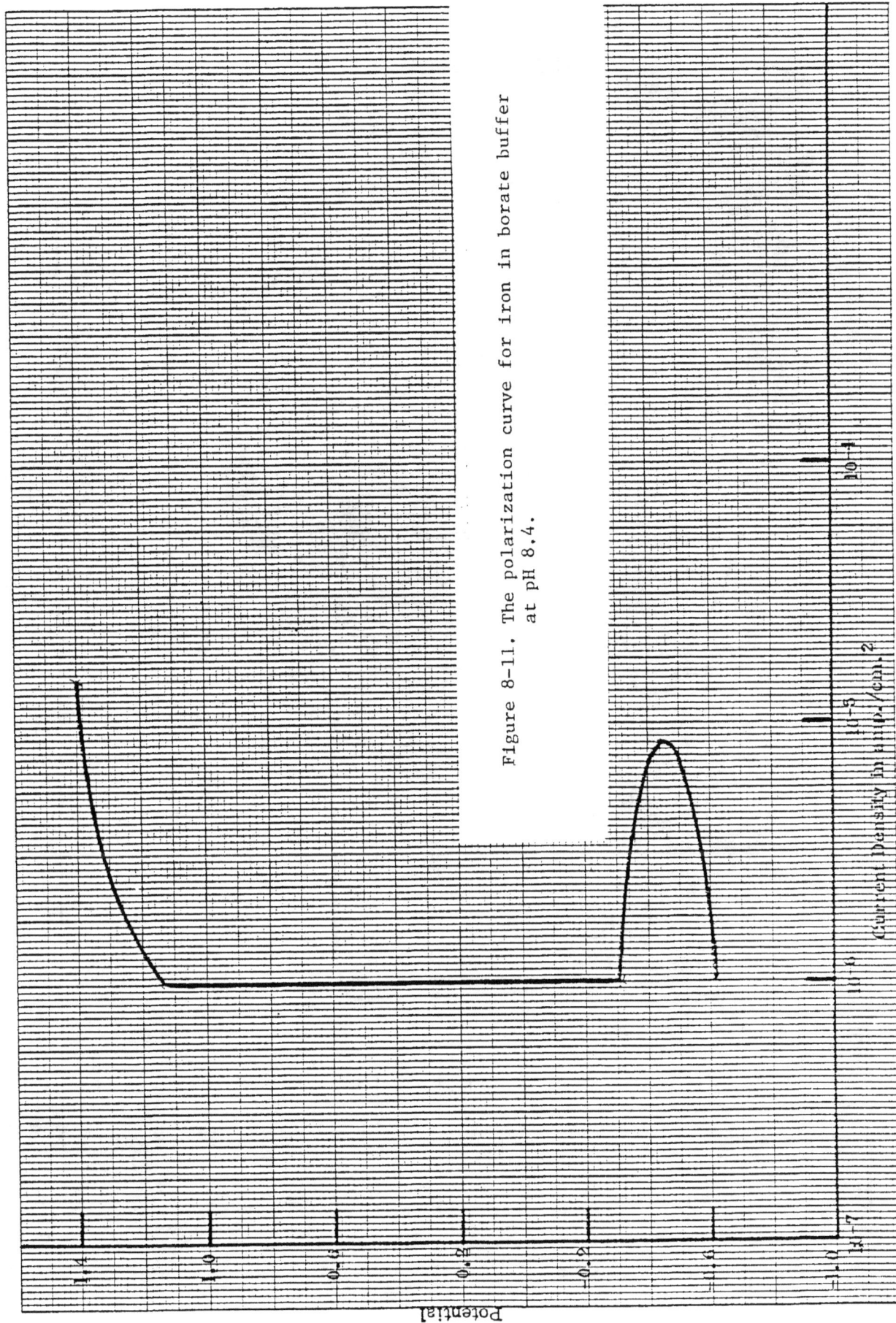

Figure 8-11. The polarization curve for iron in borate buffer at pH 8.4.

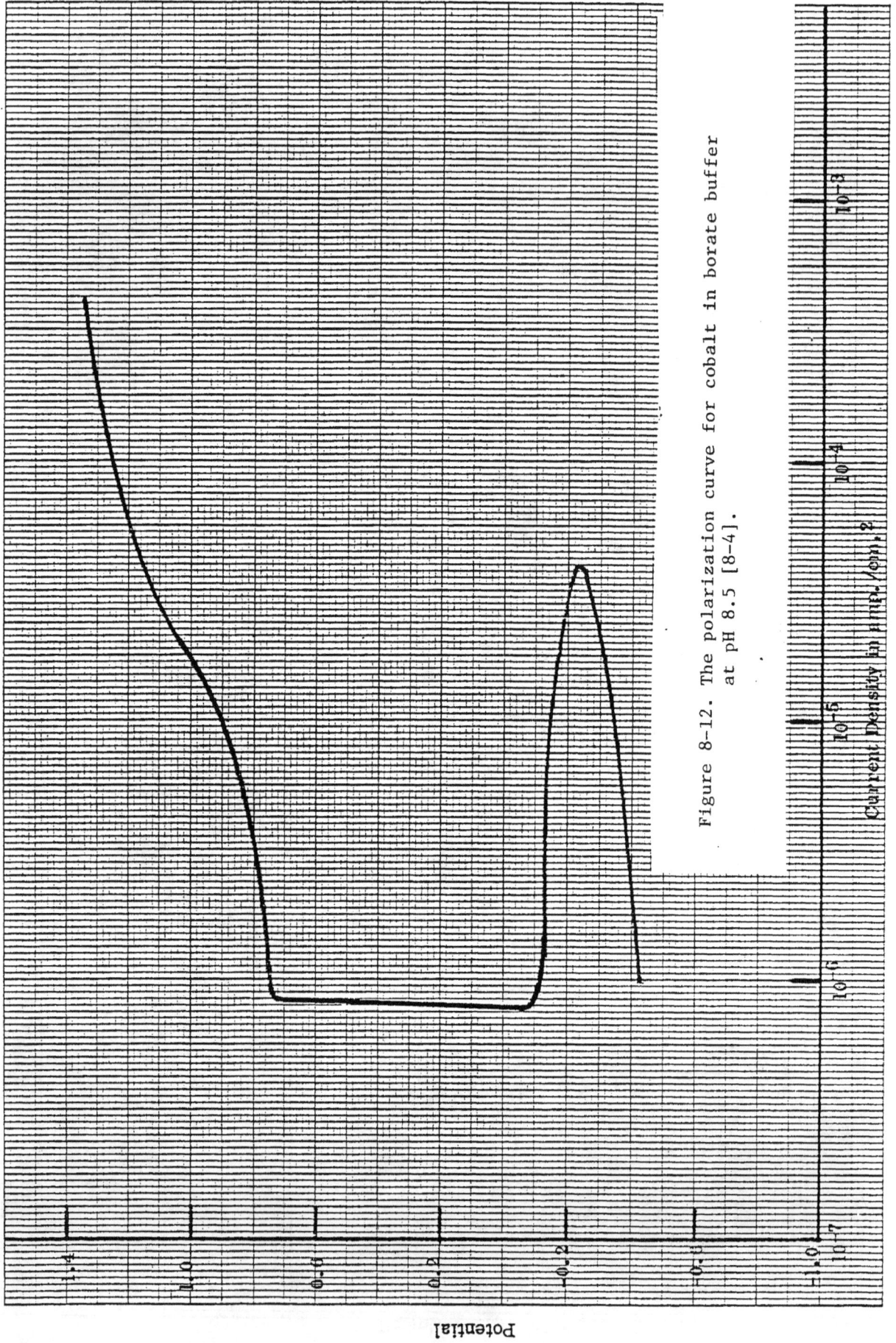

Figure 8-12. The polarization curve for cobalt in borate buffer at pH 8.5 [8-4].

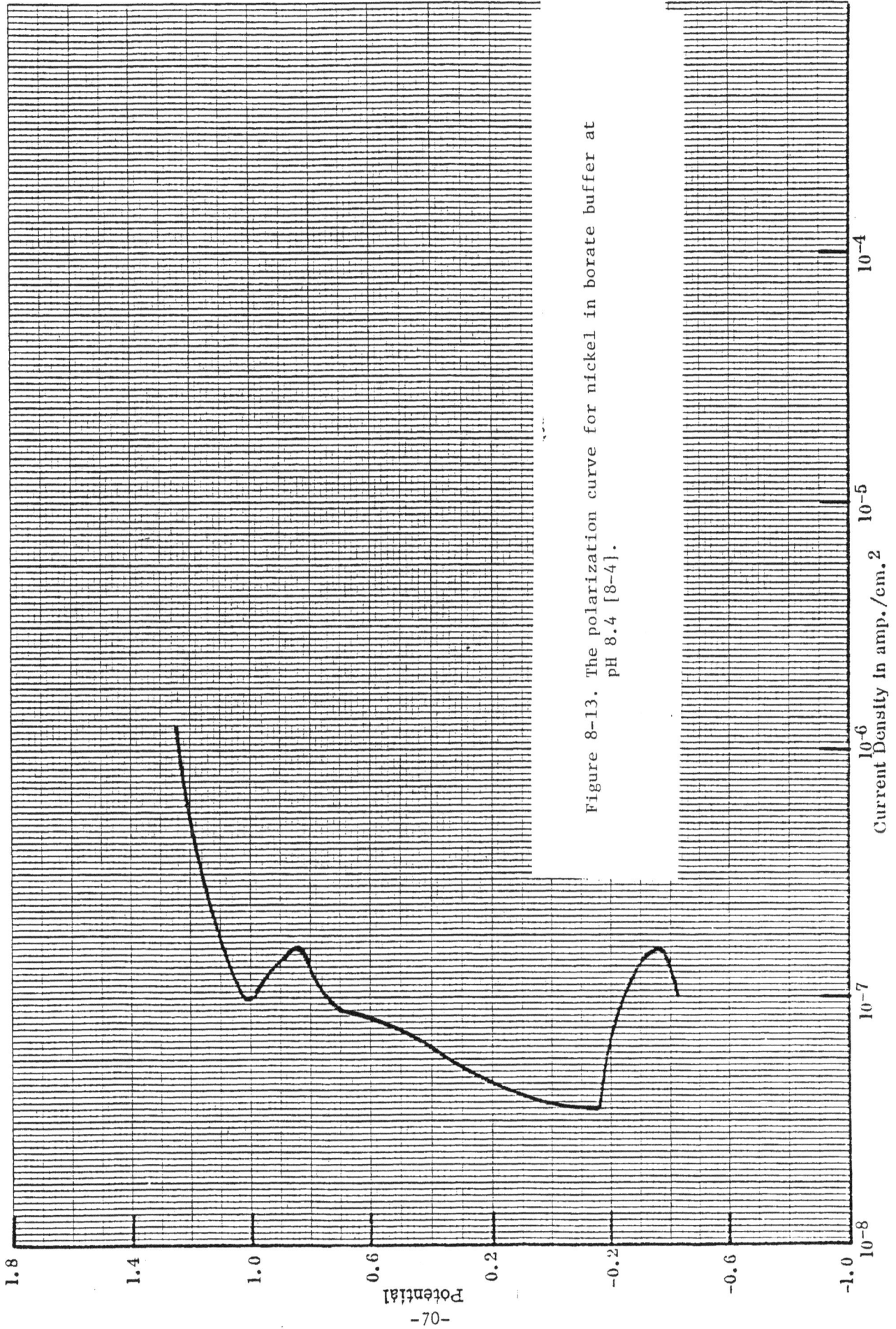

Figure 8-13. The polarization curve for nickel in borate buffer at pH 8.4 [8-4].

Table 8-1

Representative Values for Passive Current Flowing at Room Temperature in Metal/Environment Systems in Which Passivity is Observed

Metal	Environment	Current Flowing in Passive Range after Approx. 1 Hr		Reference
Iron	Deaerated 1N H_2SO_4	10^{-5} amp/cm^2	(+1.0 V)	[8-4]
Iron	Deaerated borate buffer	10^{-7}	(+0.2 V)	[8-5]
Iron	Deaerated borate buffer	10^{-8}	(-0.06 V)	[8-6]
Cobalt	Deaerated borate buffer	10^{-7}	(+0.8 V)	[8-4]
Nickel	Deaerated 1N H_2SO_4	10^{-6}	(+0.8 V)	[8-4]
Chromium	N_2 satd. 5% H_2SO_4	10^{-7}	(+0.46 V)	[8-7]
304 SS	N_2 satd. 5% H_2SO_4	10^{-6}	(+0.46 V)	[8-7]

Nagayama and Cohen [8-2] is given in Figure 8-10. Passivity commences about -0.3 V and extends up to about +1.1 V, thus covering a range in excess of 1.4 V. Above +1.1 V oxygen evolution occurs and the current rises sharply. No change occurs in the character of the surface until transpassive behavior occurs at about +1.6 V. Iron dissolves into solution as ferrous ions at potentials more negative than +0.04 but no dissolution of iron is detected when the iron is maintained at potentials in the range, +0.04 to +1.1 V. Examination of the optical characteristics of reflected plane polarized light allows one to determine the optical properties of the surface by a technique known as ellipsometry. The ellipsometric data [8-5] permit one to determine that the surface is covered with a thin film having optical properties similar (but not identical) to those of Fe_3O_4. The thickness of this oxide after polarization for 1 hour at different potentials is shown in Figure 8-14.

The passive film begins to form at the maximum anodic current (point D in Figure 8-9). At 0.0 V in borate buffer, the film on iron is estimated to be approximately 1 nm thick, although some investigators have found smaller values. When a film formed in the passive region is electrolytically reduced, a reduction curve similar to that shown in Figure 8-15 is obtained. The two waves in the electrolytic reduction curve plus many other types of studies have led many investigators to believe that the passive film has a duplex structure consisting of an inner anhydrous oxide layer represented by γ-Fe_2O_3 and an outer hydrated layer. (In the case of cobalt anodized in borate buffer, the composition of the oxide on the

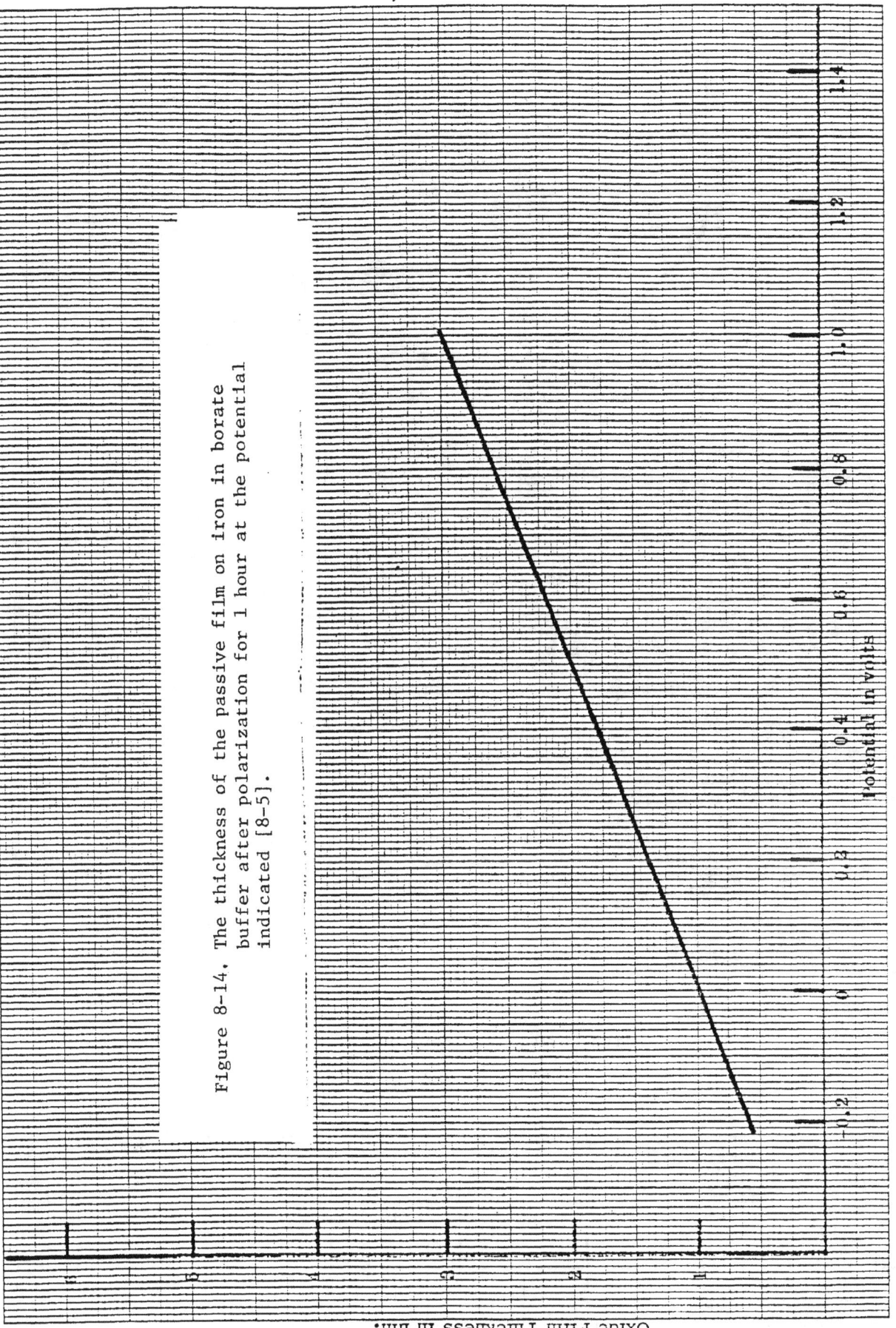

Figure 8-14. The thickness of the passive film on iron in borate buffer after polarization for 1 hour at the potential indicated [8-5].

-72-

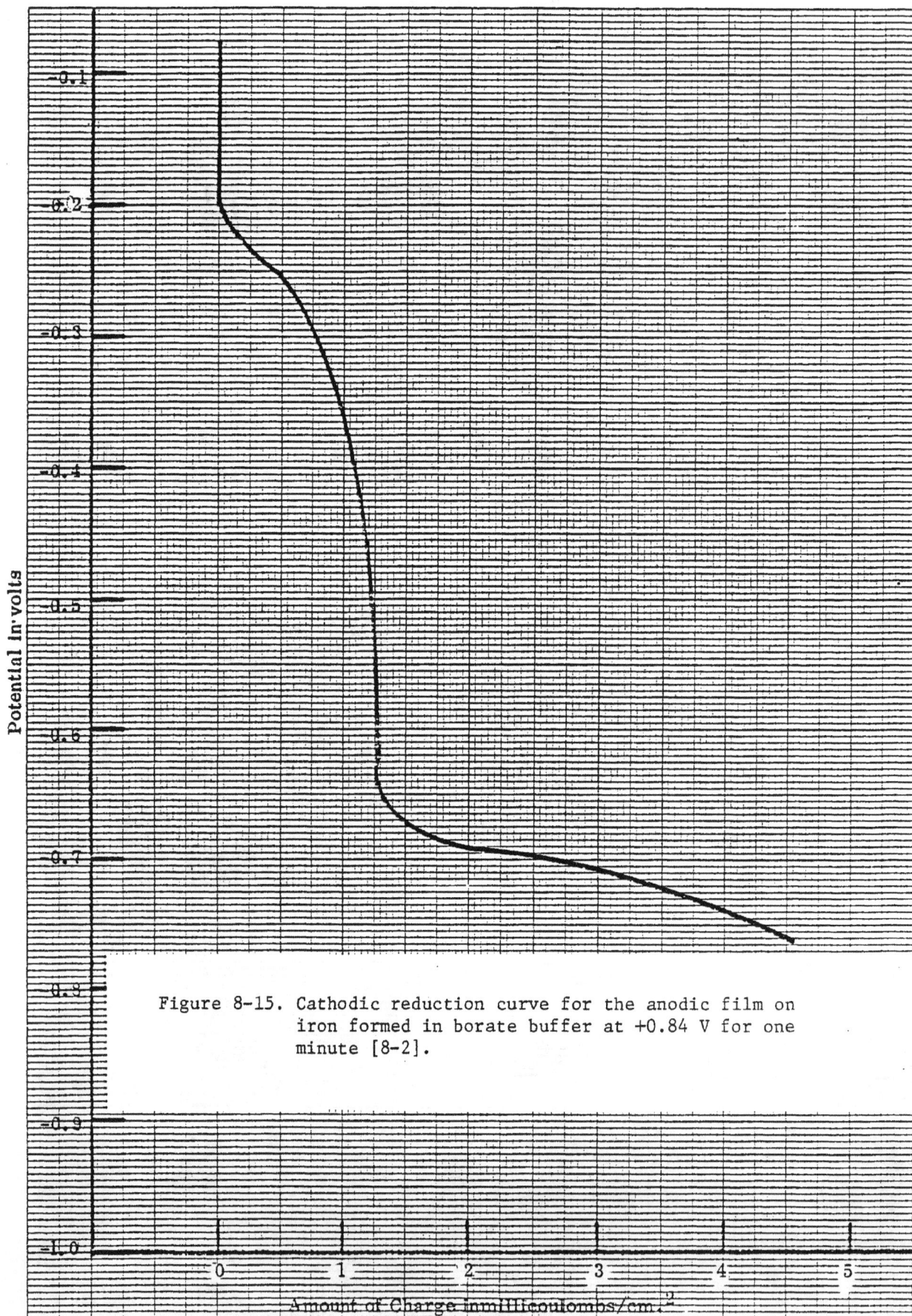

Figure 8-15. Cathodic reduction curve for the anodic film on iron formed in borate buffer at +0.84 V for one minute [8-2].

Potential in volts

Amount of Charge in millicoulombs/cm.²

surface is a function of the passive polarizing potential used [8-3]. Other workers believe the inner layer is best represented by the formula Fe_3O_4 and the other layer by γ-Fe_2O_3. There appears to be a consensus that the portion of the thin film near the metal surface has different properties than the portion of the film adjoining the solution phase.

Passivity can be induced in a metal that exhibits a normal passivity curve in the appropriate potential range by coupling it to another metal that is very active for the cathodic half-reaction. If the cathodic re-action occurs sufficiently rapidly that the anodic current exceeds the critical current for passivity, passivity results. An outstanding example of this phenomenon is shown in Figure 8-16 in which samples of titanium were intimately coupled to a second metal in boiling 2M HCl [8-8]. The potentials of these couples were such that they spanned the active/passive transition for titanium in boiling 2M HCl. Passivity of titanium was achieved when it was coupled to platinum, iridium or rhodium. Figure 8-17 describes in a schematic way the explanation for the beneficial effect of platinum when in electrical contact with titanium metal in boiling 2M HCl. The high exchange current for the reaction $2H^+ + 2e^- = H_2$ on platinum leads to the intersection of the anodic and cathodic polarization curves in the passive region of the anodic curve for titanium. This beneficial effect of the platinum metals on the rate of corrosion of titanium in hot mineral acids has been taken advantage of commercially by the development of titanium-platinum and titanium-palladium alloys. An example of the corro-sion behavior of these alloys is shown in Table 8-2.

Table 8-2

The Corrosion Rate of Titanium-Platinum Metal Alloys
in Boiling Mineral Acids [8-9]

	Corrosion Rate in mpy			
Alloy Composition	1% H_2SO_4	10% H_2SO_4	3% HCl	10% HCl
Unalloyed Ti	460	3950	242	4500
Ti-0.5% Pt	2	48	3	120
Ti-0.4% Pd	2	45	2	67
Ti-0.5% Rh	3	48	2	55
Ti-0.6% Ir	2	45	3	88

The oxide films formed on aluminum and tantalum during anodic treatment are different from the passive films formed on iron in borate solution, for example, in that the oxide is sufficiently non-conductive that it will toler-ate potential fields of the order of 10^7 volts/cm. The net effect of this

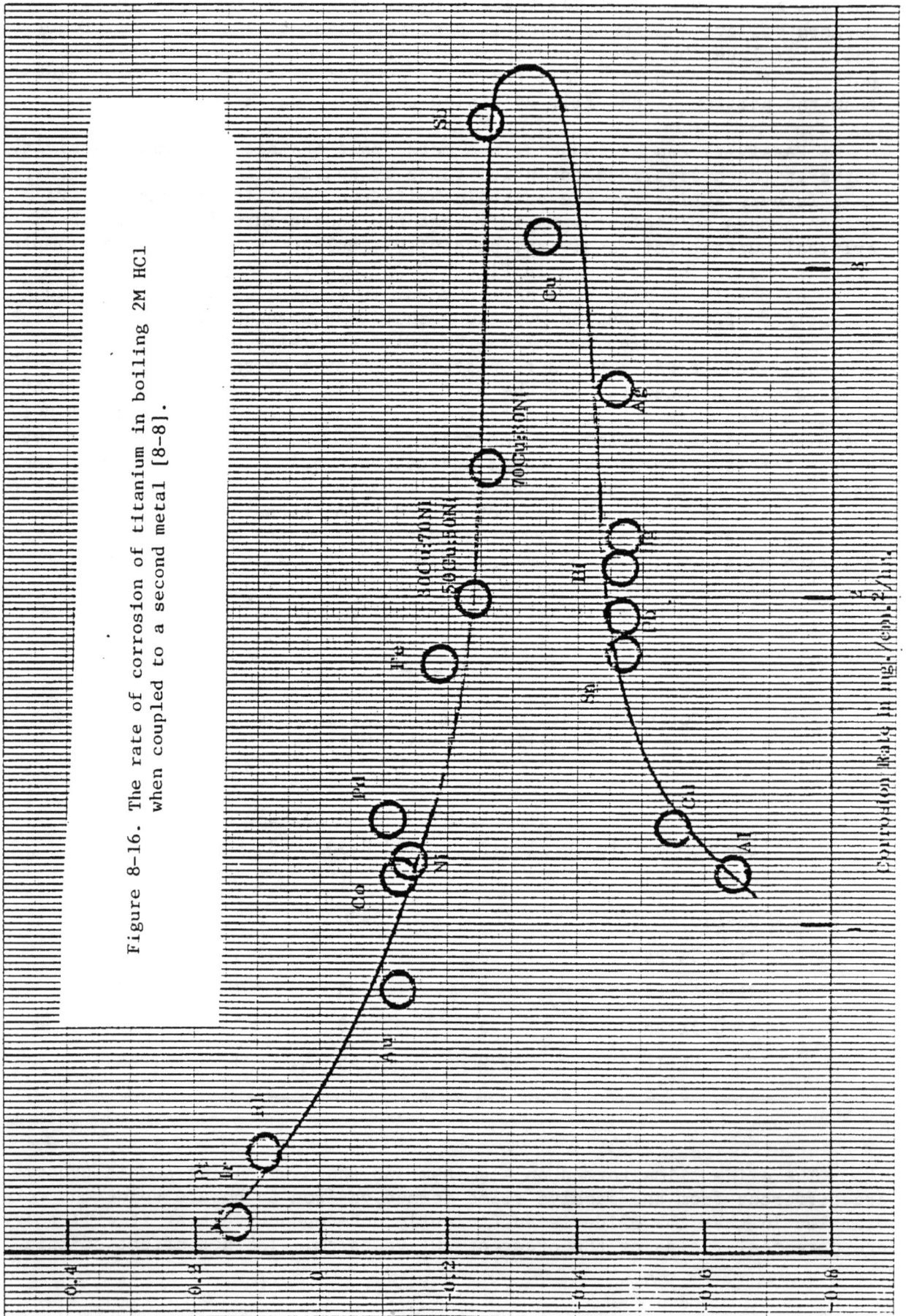

Figure 8-16. The rate of corrosion of titanium in boiling 2M HCl when coupled to a second metal [8-8].

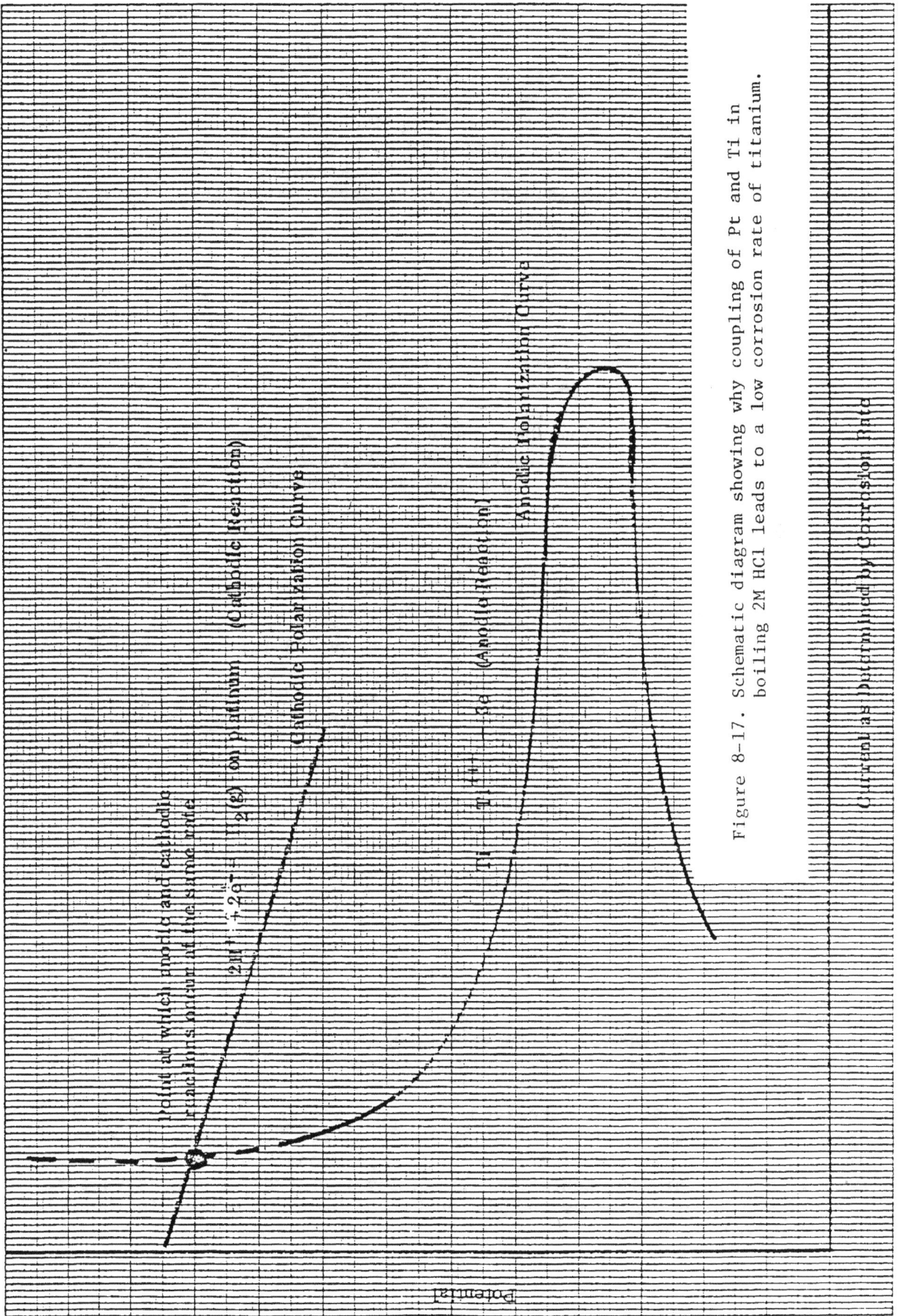

Figure 8-17. Schematic diagram showing why coupling of Pt and Ti in boiling 2M HCl leads to a low corrosion rate of titanium.

Cathodic Polarization Curve

Anodic Polarization Curve

Point at which anodic and cathodic reactions occur at the same rate

$2H^+ + 2e^- = H_2(g)$ on platinum (Cathodic Reaction)

$Ti = Ti^{+++} + 3e$ (Anodic Reaction)

Potential

Current as Determined by Corrosion Rate

property of the oxide is that the oxide film will grow when the applied field exceeds the value for the specific electrolyte and other conditions under which the anodization is carried out. Anodization of aluminum in 2M H_3BO_3 at 90° yields the curve shown in Figure 8-18, in which the coating weight (or thickness) is linearly related to the applied voltage. The non-porous oxide films formed on aluminum in electrolytes such as boric acid are not generally referred to as passive films because the thickness is of the order of 100-200 nm, and the applied potentials are approximately 100 volts. They are termed anodic films or anodic coatings

The more applied aspects of passivity occur in the case of the stainless steels and much work on passivity has been carried out on Fe-Cr, Fe-Ni, and Fe-Cr-Ni alloys. The critical current density to obtain passivity decreases as the chromium content is increased in Fe-Cr-Ni alloys as shown, for example, in Table 8-3. The oxide film formed on the surface of Fe-Cr and Fe-Cr-Ni alloys is enriched in chromium as determined by Auger electron spectroscopy and ion scattering spectroscopy. This enrichment increases with increase in chromium content of the substrate, time held at passivating potential, and as the polarizing potential is made more noble [8-12].

Table 8-3

Critical Current Density for Passivity in 1.28N H_2SO_4
at 25° in Fe-Ni-Cr Alloys [8-11]

Alloy Composition	Critical Current Density for Passivity
20% Ni - 3% Cr	300 mamp/cm^2
20% Ni - 5% Cr	200
20% Ni - 8% Cr	50
20% Ni - 10% Cr	2
20% Ni - 17% Cr	1

At a passivity conference in Hawaii in 1975, there was general agreement among participants that the composition of the passive film on iron-chromium alloys changes as the chromium content increases. At higher chromium contents the amount of iron in the passive film decreases. In passive films on pure iron, the valence state of the iron is Fe^{+++}, whereas in the passive film on Fe-Cr alloys, the iron exists both as Fe^{++} and Fe^{+++}. The chromium content in the passive film is greater than the proportional chromium content in the metal. As the chromium content in the alloy is increased, the structure of the oxide becomes amorphous and the number of hydroxyl groups in the film becomes greater.

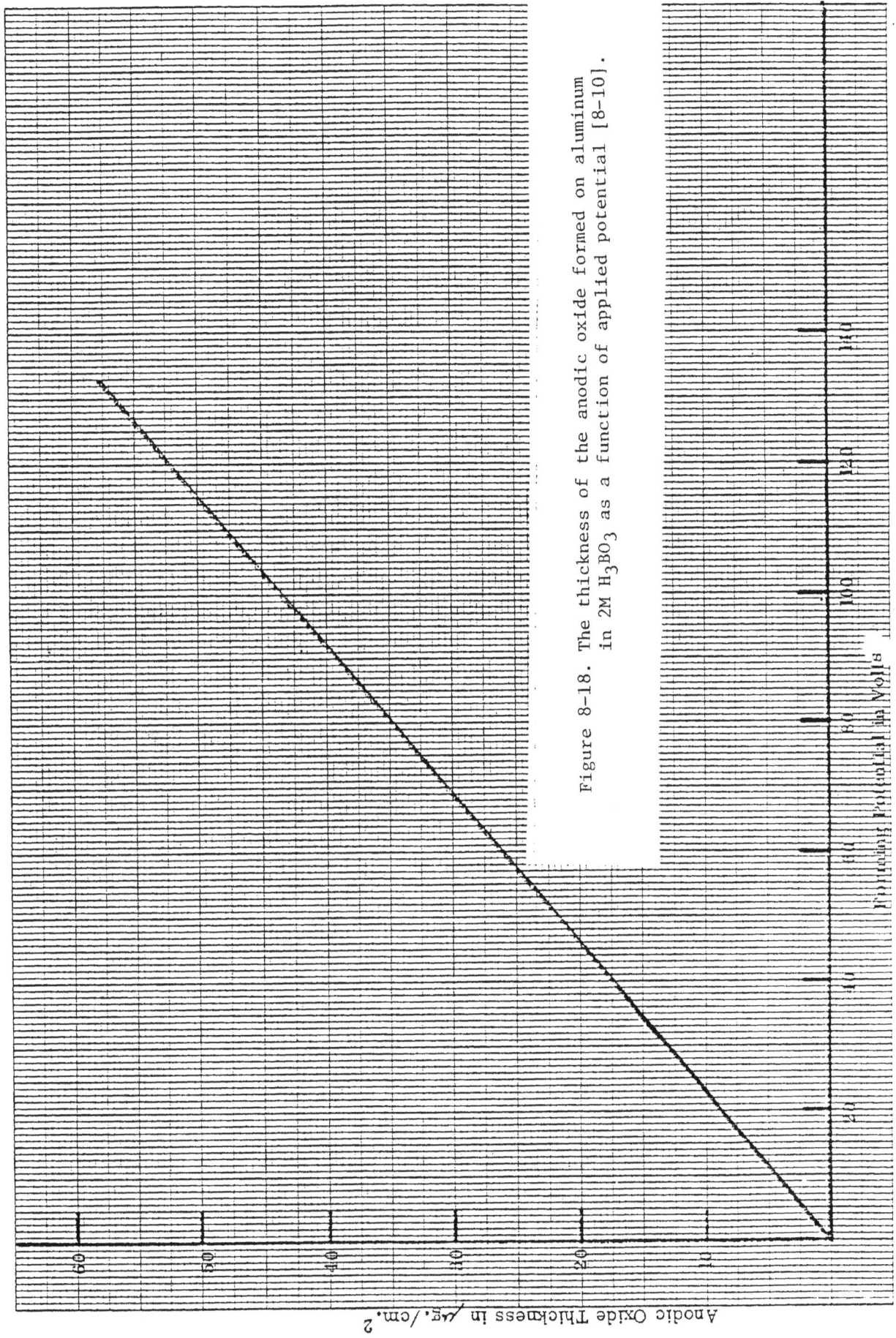

Figure 8-18. The thickness of the anodic oxide formed on aluminum in 2M H_3BO_3 as a function of applied potential [8-10].

Chemical Destruction of Passivity

The chemical breakdown of passivity leads to many destructive forms of corrosion, such as pitting, crevice corrosion, and intergranular corrosion. The following comments will be concerned with the breakdown of passivity and will not discuss the consequences of the breakdown. Chemical breakdown is concerned with changes in the chemistry of the passive film or in the environment that result in alterations in the properties of the passive film such that the anodic reaction, $M - ne^- = M^{+n}(aq)$, can occur at a significant rate.

Several generalizations may be made about the chemical breakdown of passivity. Any model to explain the events must take these facts into account:

1. Breakdown occurs above a certain rather well-defined potential that is a function of the temperature and environmental composition.

2. Certain species, such as the monovalent Cl^-, Br^-, and I^- ions, are particularly effective in causing breakdown. The F^- ion is especially effective in destroying the passivity of titanium.

3. Breakdown generally occurs in a highly localized fashion.

4. When aggressive ions, such as Cl^-, are introduced into a medium in which the metal is exhibiting passive behavior, there is an induction time before localized attack occurs.

Many models have been proposed to account for these facts but the model that seems to be consistent with all information is the following: Metal cations are removed from the lattice of the passive oxide film when chloride ions penetrate into the oxide and become the coordinate ions surrounding the metal cation. In the case of aluminum, it may require 4 Cl^- so as to form the soluble complex $AlCl_4^-$ or 3 Cl^- ions to form a complex such as $[AlCl_3OH]^-$. The critical potential is that potential at which a sufficient number of Cl^- ions are at the oxide/solution interface to have a finite probability that the necessary number of Cl^- ions will be present around an aluminum cation. The potential at the oxide/solution interface determines the composition of ions on the oxide side of the double layer. The more polarizable ions tend to dominate as the potential is made more noble but the concentration in solution is also a determining factor. When other anions are present in solution they compete with Cl^- ions for surface sites.

The aggressive ions such as Cl^-, Br^-, I^-, and F^- are monovalent and, when they become incorporated in the oxide by diffusion from the adsorbed state, they replace the divalent oxygen ion, resulting in defects in the oxide with consequent more rapid migration of ions in the lattice. Ambrose and Kruger [8-13] have shown that incorporation of Cl^- ions in the oxide film on iron accelerates the rate of breakdown.

The localized nature of breakdown may be dependent upon pre-existing defects that serve as loci for attack, or the breakdown may be purely on a statistical basis. Either consideration may dominate depending on the purity of the metal and the conditions under which the passive film was formed.

Once attack has started and a single cation has been removed, there exists a vacancy nucleus that makes easier the likelihood for Cl^- penetration and geometric positioning around a second Al^{+++} cation. The stepwise removal of adjoining cations in the film suggests that the incubation period before attack of the substrate metal occurs should increase with increase in thickness of the passive film. McBee and Kruger [8-14] have indeed shown that the time to breakdown of the passive film on iron in borate buffer at pH 8.4 on addition of 0.005N NaCl increased with increase in film thickness as detailed in Figure 8-19. Evidence also exists from a number of studies showing that annealing of the oxide film to reduce the number of structural defects increased the incubation time for breakdown.

Effect of Alloy Composition on Breakdown of Passivity

One of the techniques used to determine when breakdown of passivity occurs is the tendency of the alloy to pit and much of the knowledge of the effect of alloy composition on breakdown comes from pitting studies.

The susceptibility of iron to pitting can be greatly changed by alloying. The two more effective elements in reducing pitting attack are chromium and molybdenum. Silicon, vanadium and rhenium are also beneficial in reducing pitting attack on an 18-14 Cr-Ni steel [8-15]. Conflicting reports on the effectiveness of titanium additions are recorded. These results may be a consequence of titanium's ability to bind carbon and nitrogen and thus prevent local reduction of the chromium concentration by carbide or nitride formation. High concentrations of titanium result in the formation of second phases that may serve as sites for pit initiation.

Molybdenum additions to high purity 25% chromium steels in the amounts of 3.5-5% result in alloys having very high resistance to pitting and crevice corrosion in 1N HCl and 0.33M $FeCl_3$. Alloys containing commercial amounts of impurities are not immune to pitting but show decreased tendency to pitting in comparison with alloys devoid of molybdenum [8-16]. Molybdenum additions greatly decrease the critical current density necessary for passivity as shown, for example, in Figure 8-20 for 25% Cr steels in 1N HCl. Molybdenum also changes significantly the critical potential for passivity as shown in Figure 8-21.

The reasons for the very beneficial effects of molybdenum on the resistance to pitting of chromium steels are still unknown. Molybdenum increases the resistance to chemical breakdown of the passive film but Auger analysis of passive films on stainless steels containing significant amounts of molybdenum have not been able to show the presence of molybdenum in the film. The most likely explanation is that molybdenum alters the defect character of the passive film and thereby decreases the ability of chloride to penetrate into the film.

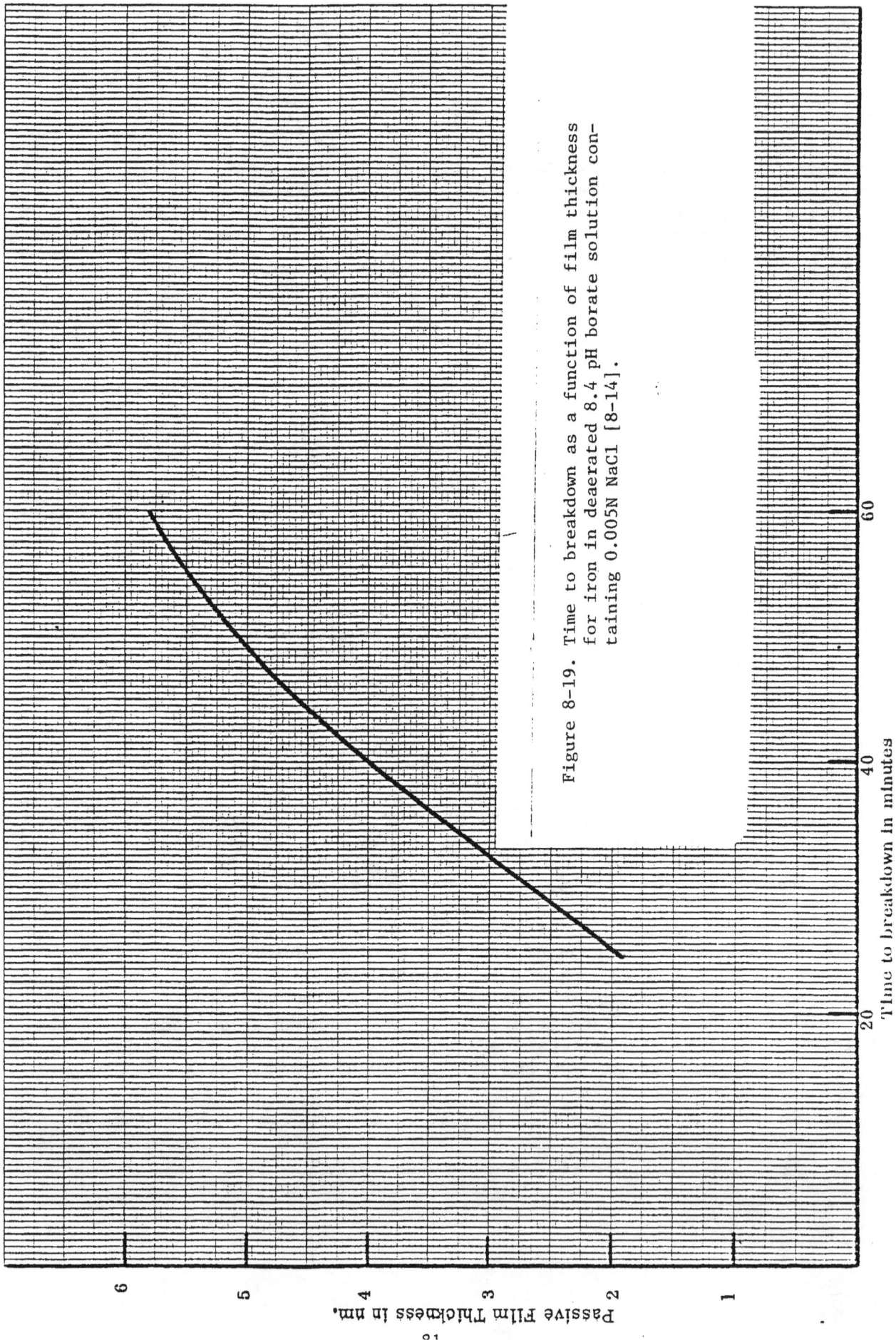

Figure 8-19. Time to breakdown as a function of film thickness for iron in deaerated 8.4 pH borate solution containing 0.005N NaCl [8-14].

Time to breakdown in minutes

Passive Film Thickness in nm.

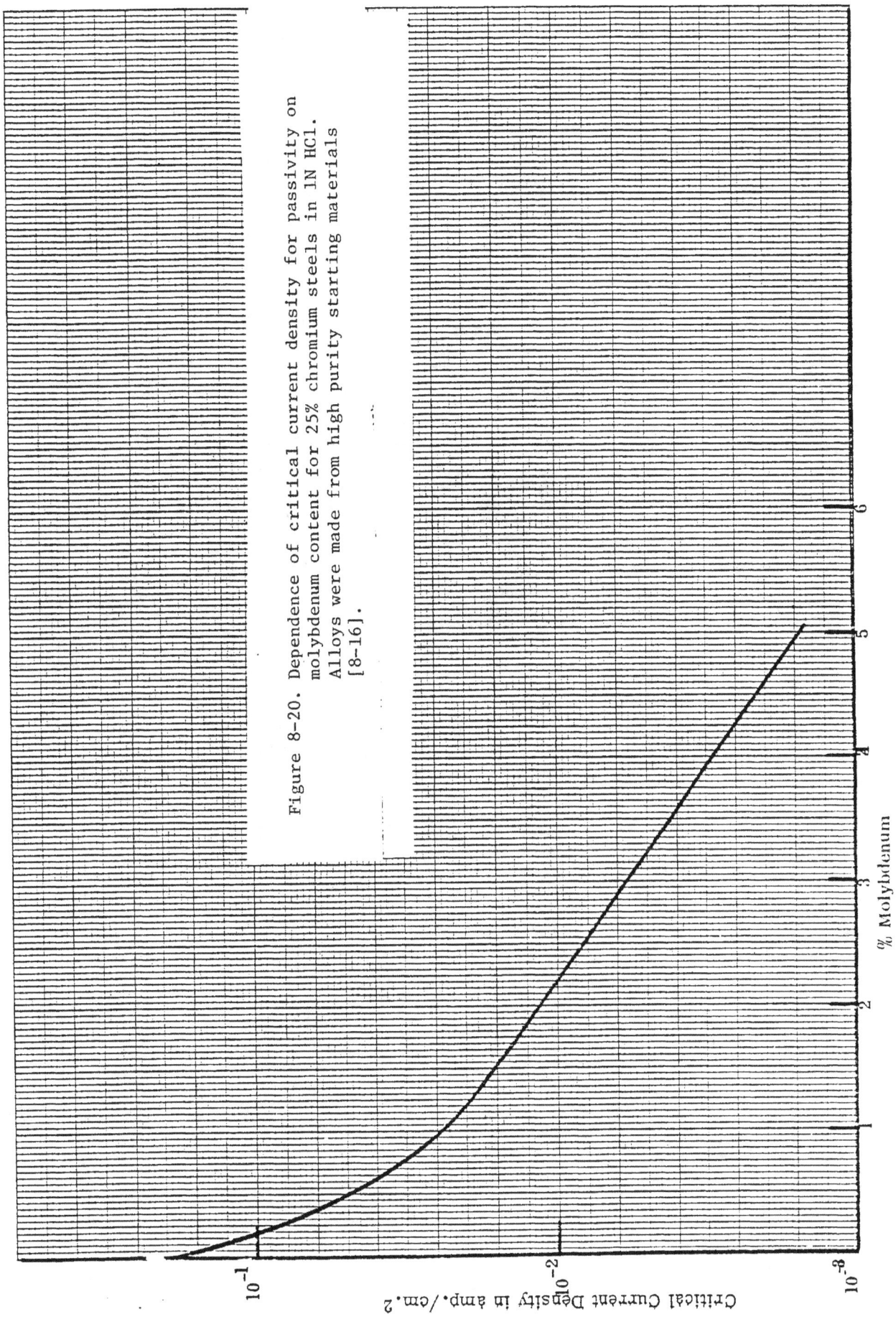

Figure 8-20. Dependence of critical current density for passivity on molybdenum content for 25% chromium steels in 1N HC1. Alloys were made from high purity starting materials [8-16].

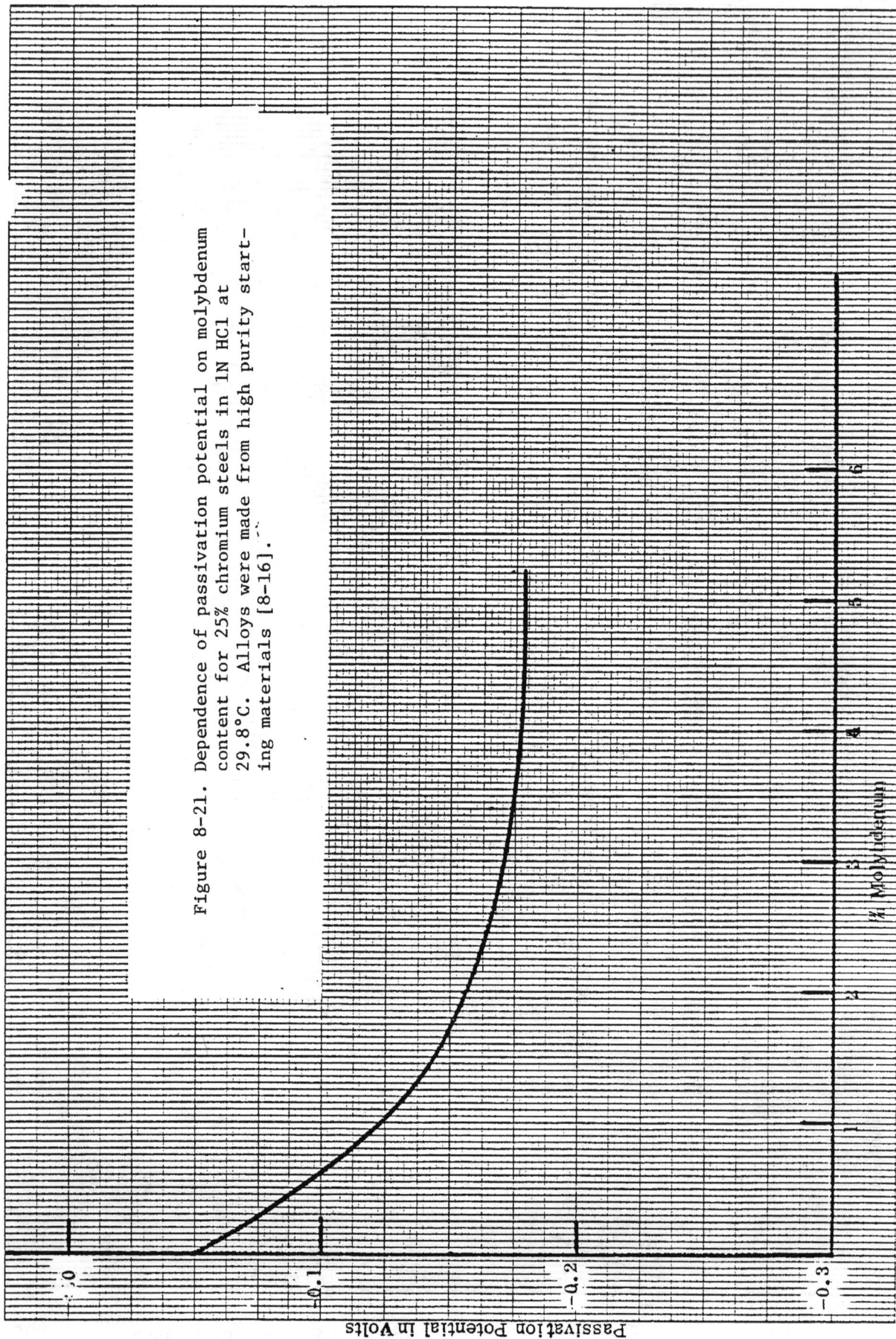

Figure 8-21. Dependence of passivation potential on molybdenum content for 25% chromium steels in 1N HCl at 29.8°C. Alloys were made from high purity starting materials [8-16].

Mechanical Destruction of Passivity

Since the passive character of a metal is determined by an oxide present on the surface, the passive nature should be destroyed when the oxide is mechanically removed or fractured in any way. In the case of a metal such as aluminum, the passive film is immediately repaired by reaction with water and any temporary reactive state is lost. The reality of the passive oxide on the surface of aluminum and the high reactivity of the aluminum is readily revealed by a technique known as strain electrometry in which a metal, whose potential versus a reference is monitored continuously, is suddenly strained. A typical example of this phenomenon is shown in Figure 8-22. An aluminum wire with a steady-state potential of -0.42 V while immersed in 0.1M NaCl at pH of 7.5 was abruptly strained 4.5%. The oxide film on the surface was ruptured by this sudden elongation and the active aluminum was exposed to the electrolyte. The potential of the aluminum changed to -1.46 V in 0.04 sec and rapidly returned to its steady-state value in slightly more than 1 sec. This negative potential of -1.46 V is indicative of a very active aluminum surface.

The experimental results are schematically explained in Figure 8-23. Although the apparent surface area is not greatly increased by straining, the active surface on which the anodic reaction, $Al - 3e^- = Al^{+++}$, is occurring increased by approximately 19 orders of magnitude, assuming that the cathodic polarization curve has a slope of 0.059 V/decade. It should be emphasized that the horizontal axis in the figure represents current, not current density as normally drawn. The cathodic curve is not shifted by the straining since it is assumed that the cathodic reaction takes place uniformly over the surface and the straining did not significantly change the surface area available for the cathodic reaction. Curve I that intersects at point A represents the situation before straining — the current is very low and the corrosion rate is essentially negligible. Immediately after straining, the anodic reaction is represented by Curve II and the corrosion current is represented by the intersection point B. As the oxide film is restored on the active surface by reaction with water, the potential moves up along the cathodic curve as the active surface area is decreased and, after approximately 1 sec, returns to the steady state value at point A.

There is much interest in the kinetics of repair of an oxide film on a metal after it is damaged by abrasion, scratching, etc. The subject is of special interest to those who are concerned with stress corrosion cracking and the phenomena that occur after a crack is nucleated at the surface. Ambrose and Kruger [8-18] have pioneered a technique they term "tribo-ellipsometry" in which a surface is abraded while immersed in a fluid and the rate of regrowth of the oxide on the surface is followed by a short-response-time ellipsometer. An example of the data obtained from this technique is shown in Figure 8-24. The rate of film growth on a steel specimen after abrasion occurred much more rapidly in a solution of 1N $NaNO_2$ than in a solution of 1N $NaNO_3$. Although the technique is in its infancy and much more work must be done before conclusions can be drawn, it is interesting to point out that this steel is susceptible to stress corrosion cracking in $NaNO_3$ solution and is not susceptible in $NaNO_2$ solution.

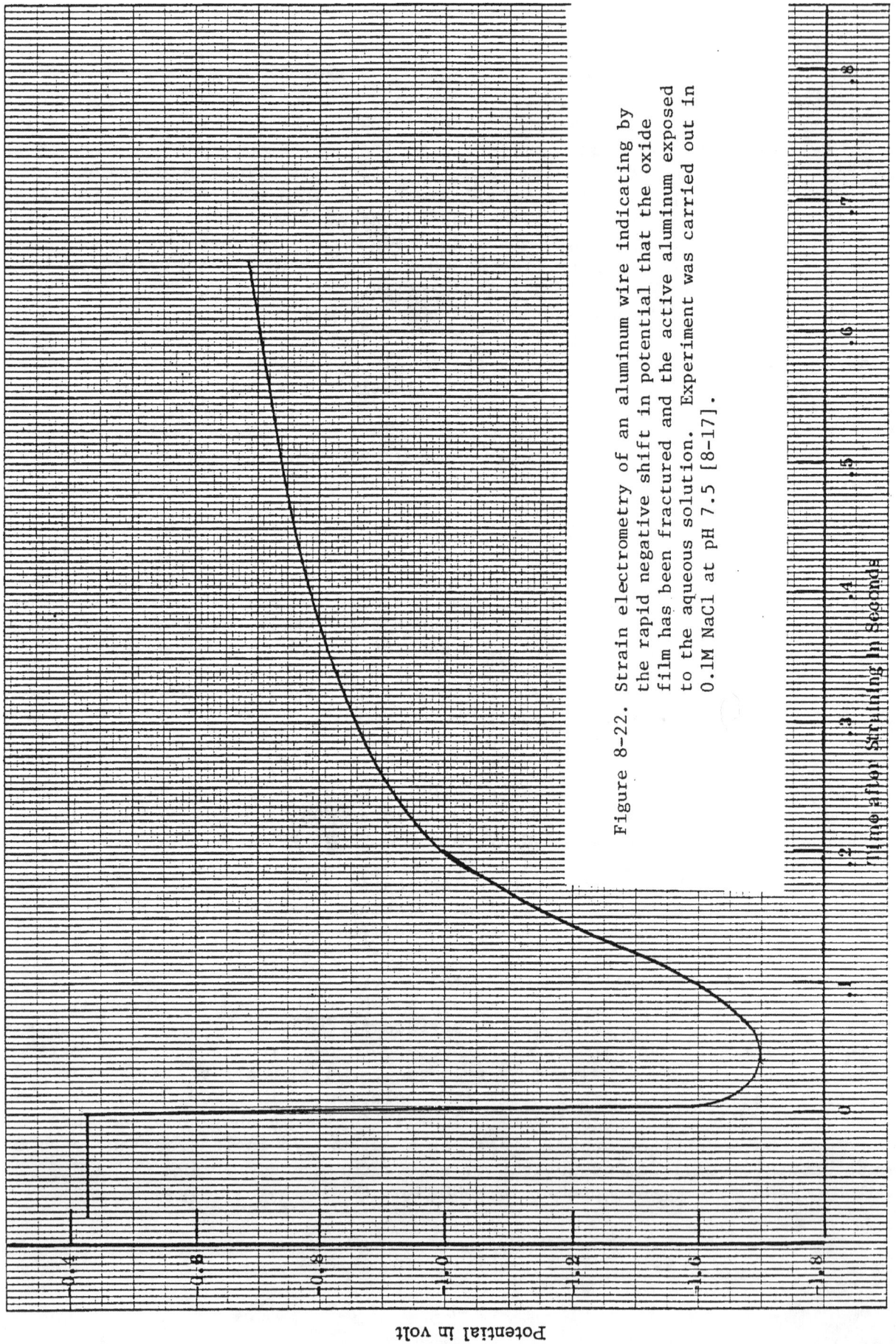

Figure 8-22. Strain electrometry of an aluminum wire indicating by the rapid negative shift in potential that the oxide film has been fractured and the active aluminum exposed to the aqueous solution. Experiment was carried out in 0.1M NaCl at pH 7.5 [8-17].

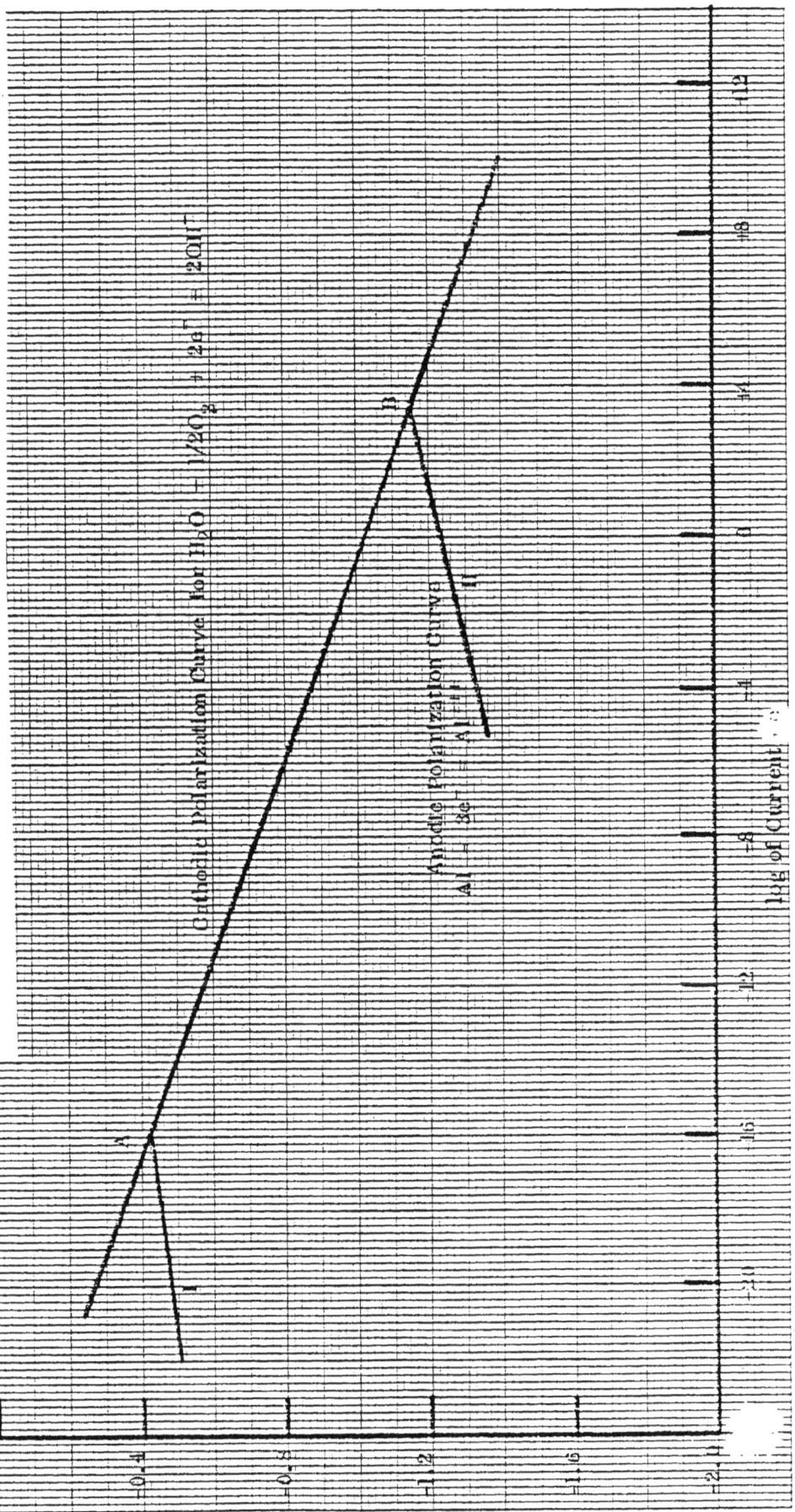

Figure 8-23. Schematic explanation for the potential behavior of aluminum during straining as observed experimentally in Figure 8-13.

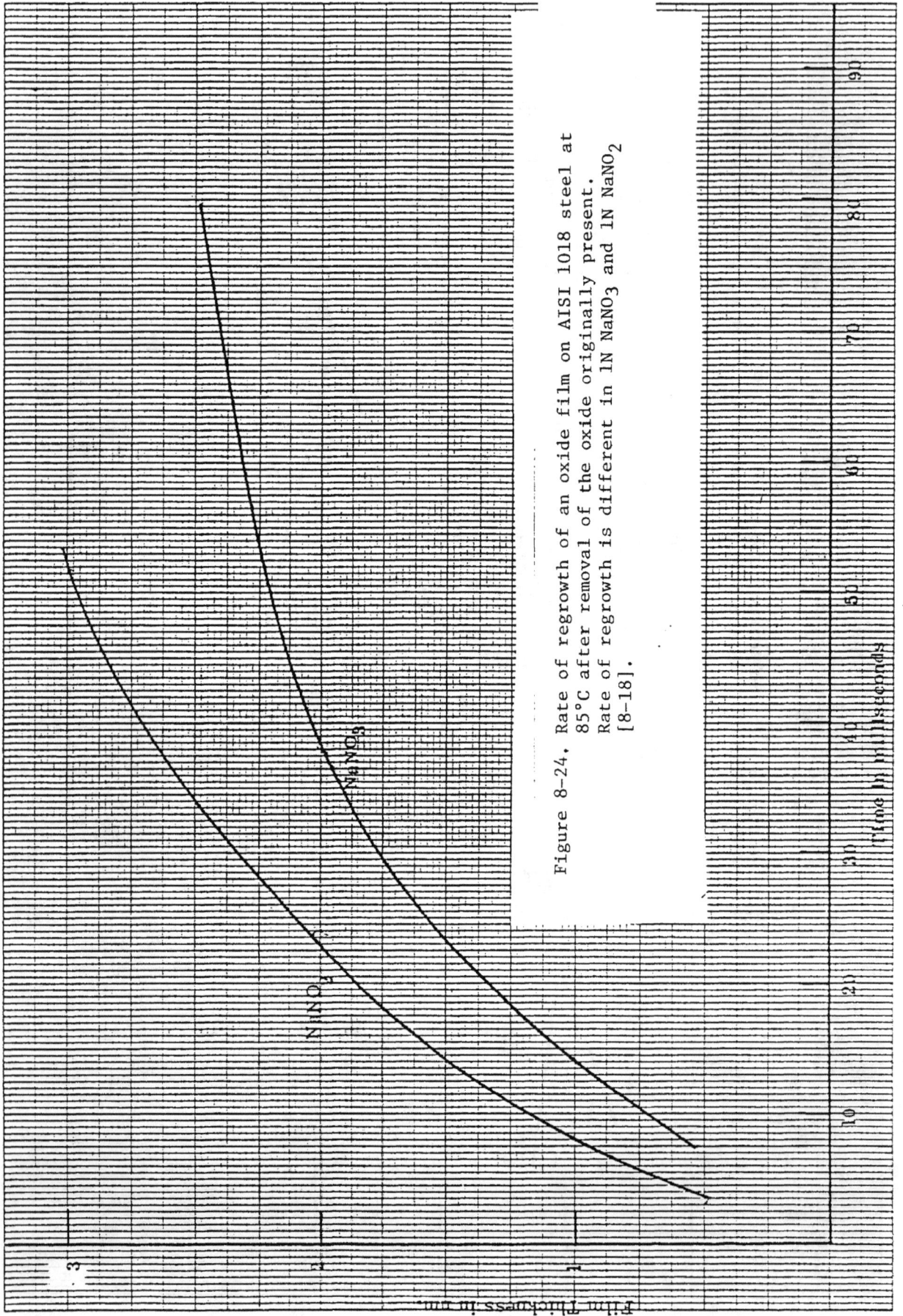

Figure 8-24. Rate of regrowth of an oxide film on AISI 1018 steel at 85°C after removal of the oxide originally present. Rate of regrowth is different in 1N NaNO$_3$ and 1N NaNO$_2$ [8-18].

References

[8-1] See Trans. Faraday Soc. $\underline{9}$, 203-90 (1914).

[8-2] M. Nagayama and M. Cohen, J. Electrochem. Soc. $\underline{109}$, 781 (1962).

[8-3] G. W. Simmons, E. Kellerman, and H. Leidheiser, Jr., J. Electrochem. Soc. $\underline{123}$. 1276 (1976).

[8-4] N. Sato, "Passivity and Its Breakdown on Iron and Iron Base Alloys," edited by R. W. Staehle and H. Okada, Natl. Assocn. Corrosion Engrs., Houston, Texas, 1976, p.1.

[8-5] N. Sato and K. Kudo, Electrochim. Acta $\underline{16}$, 447 (1971).

[8-6] R. P. Frankenthal, Electrochim. Acta $\underline{16}$, 1845 (1971).

[8-7] T. Ishikawa, "Passivity and Its Breakdown on Iron and Iron Base Alloys," edited by R. W. Staehle and H. Okada, Natl. Assocn. Corrosion Engrs., Houston, Texas, 1976, p.49.

[8-8] W. R. Buck III, B. W. Sloope, and H. Leidheiser, Jr., Corrosion $\underline{15}$, 566t (1959).

[8-9] M. Stern and H. Wissenberg, J. Electrochem. Soc. $\underline{106}$, 759 (1959).

[8-10] G. A. Dorsey, Jr., J. Electrochem. Soc. $\underline{116}$, 466 (1969).

[8-11] H. H. Uhlig, "Passivity and Its Breakdown on Iron and Iron Base Alloys," edited by R. W. Staehle and H. Okada, Natl. Assocn. Corrosion Engrs., Houston, Texas, 1976, p.21.

[8-12] H. Okada, H. Ogawa, I. Igoh, and H. Omata, "Passivity and Its Breakdown on Iron and Iron Base Alloys," edited by R. W. Staehle and H. Okada, Natl. Assocn. Corrosion Engrs., Houston, Texas, 1976, p.82.

[8-13] J. R. Ambrose and J. Kruger, Proc. 4th International Congress on Metallic Corrosion, 698 (1972).

[8-14] C. L. McBee and J. Kruger, "Localized Corrosion," edited by R. W. Staehle, B. F. Brown, J. Kruger, and A. Agrawal, Natl. Assocn. Corrosion Engrs., Houston, Texas, 1974, p.252.

[8-15] N. D. Tomashov, O. P. Chernova, and N. Markova, Corrosion $\underline{20}$, 166t (1964).

[8-16] E. A. Lizlovs and A. P. Bond, J. Electrochem. Soc. $\underline{118}$, 22 (1971).

[8-17] H. Leidheiser, Jr., and E. Kellerman, Corrosion $\underline{26}$, 99 (1970).

[8-18] J. R. Ambrose and J. Kruger, Corrosion $\underline{28}$, 30 (1972).

Section 9

Metallic Coatings on Non-Metallic Substrates

Note: This section has been taken from an article by Charles
Davidoff in the 1984 edition of "Metal and Finishing
Guidebook."

Metallizing Processes

A. Conductive Paints:

The part to be metallized is coated with a conductive paint. This is a
lacquer or varnish in which is suspended a conductive pigment such as graphite,
copper or silver. After the part has been thoroughly dried it is generally
plated in a standard acid-copper bath.

Particular care must be taken to preseal wood, plaster, and other porous
parts to prevent absorption of plating solution which would gradually bleed out
over the surface of the part.

Silver paint containing 60 to 70% silver metal pigment is preferred for
brushing and lesser percentages for spraying. Cellulosic ester and methacrylate
type resins are generally used as the binders for these paints. Epoxy bases
have now become available. They are sold as proprietary products by several
large and small manufacturers under the descriptive name of "Silver Conductive
Paint."

More recently, several conductive paints with gold as the conductor and
epoxy resin as the binder have appeared on the market. These are largely
directed to use on non-conductive materials used in the electronics industry.

A copper paint, sometimes called "bronzing" paint, is often used for
application as a conducting film. The mixture consists of the following:

Nitrocellulose lacquer......1 fl oz
Lacquer thinner............7 fl oz
Copper lining powder........2 oz

Only enough for immediate use should be prepared since the metal powder
often causes the lacquer to jell. If the copper powder is greasy, it should be
washed with thinner before using. If sprayed, the copper paint should be applied with the gun held at a distance so that the film dries almost as soon as
it reaches the surface. A glossy appearance indicates that the copper is
coated with a layer of lacquer, which will prevent passage of current.

A good method of insuring that the surface is conductive is to dip the coated article in a solution of about 1 oz/gal silver cyanide and 4 oz/gal sodium cyanide. Absence of immersion silver deposit in areas will indicate that the lacquer film has not been applied correctly. The silver deposit also provides a better conducting medium than the copper for subsequent electroplating.

B. Mirroring Processes:

To serve as a base for subsequent plating requires a reasonably thick and adherent film. To serve as a pleasantly appearing polished metal part requires a "first surface" type to be precipitated on a surface prepared to give reasonable adhesion, and prelacquered to produce a lustrous and receptive base for a mirror.

For the functional metal film, which serves as a base for further plating, it is common to roughen the plastic base first to insure adhesion. This can be done mechanically by depolishing through dry abrasive blasting or wet abrasive tumbling or chemically by etching, as with solvents, oxidizing acids or caustic solutions.

These all have as their basic aim the provision of a mechanically locking surface and an hydrophilic surface (one that wets well).

An example of a chemical etchant is the use of 40 grams of sodium dichromate dissolved in 750 milliliters of sulfuric acid (66° Be) and 250 milliliters of water. (Caution: always add acid to water, and make sure mix is cold before dissolving in the sodium dichromate.) The bath is used hot (85 to 95°C) for a five-second interval. Parts so treated are rinsed and dipped again for five seconds in a 15% by weight sodium hydroxide solution maintained at 90 to 100°C. Another variation of the chemical etchant is

Chromic acid......10 oz/gal
Sulfuric acid.....32 fl oz/gal

This is used at room temperature for one to two minutes. Again, wash and neutralize surface in sodium hydroxide before proceeding. If such an etching process is used, be careful not to over-react the surface. This can deteriorate and weaken the plastic.

Pretreatment involving phosphoric and chromic acid has also been used. In the pretreatment of cellulose triacetate a pretreatment in aqua regia for about ten to twenty minutes is sometimes used. The nature of the surface imparted by these etching processes is fundamental to developing the maximum strength of metal to plastic adhesion.

The roughening is followed by cleaning and then sensitizing. This is a brief soak (one to five minutes) in stannous chloride solution.

There are many formulas indicated. The usual ones contain stannous chloride and hydrochloric acid in varying proportions. The tolerance for ratio variations is wide. Two typical formulas are as follows:

        1. Stannous chloride .... 10 g
           Hydrochloric acid .... 40 cc
           Water .... 1 liter

        2. Stannous chloride .... 180 g
           Hydrochloric acid .... 180 cc
           Water .... 200 cc

Other sensitizers have been used such as gold chloride, palladium chloride, platinum chloride, stannous fluoborate, silicon tetrachloride, and titanium tetrachloride. A recent patent suggests the use of alkali gold sulfite.

Sometimes gold chloride, platinum chloride or palladium chloride are used as an additional or second sensitizing step. When they are so used we often refer to these treatments as nucleators. A typical palladium chloride bath for this function is

        Palladium chloride .... 0.005 to 2 g/1
        Hydrochloric acid .... 0.1   to 2 g/1

Despite this apparent wide range of chemicals which will sensitize a surface to receive a mirror type film, the step is important and critical.

It is equally important to carefully wash off all traces of the sensitizer before a metal surface is precipitated. The final rinse here should be either distilled or demineralized water.

A good silver film is the type deposited by the "Brashear Formula."

        a. Silver Nitrate Solution:
           Silver nitrate .... 20 g
           Potassium hydroxide .... 10 g
           Distilled water .... 400 ml

A precipitate is formed and is just dissolved with ammonium hydroxide (approx. 50 cc).

Caution: To avoid formation of explosive fulminates, the silver salt, caustic, and ammonia should never be mixed in concenrated form, but should be diluted with water first. Containers which have held this fulminate solution should be washed carefully and never allowed to dry with any residual material. Dry, this material is explosive.

        b. Reducing Solution:
           Cane sugar .... 90 g
           Nitric acid .... 4 ml
           Distilled water .... 1 liter

        Boil and cool before using.

Immediately before using, mix one part of reducer with four parts of silver. A reaction temperature of 68°F is preferable.

Other good formulas are given in National Bureau of Standards Circular No. 389.

A copper film may also be deposited by reduction with formaldehyde of a Fehling solution to which a very small amount of silver is added. A typical example is:

Anhydrous copper sulfate	2	g
Silver nitrate	0.2	g
Rochelle salt	4	g
Potassium hydroxide	4	g
Distilled water	100	cc

Reduce with 5% formaldehyde solution.

A recent copper formula, reported by E. B. Saubestre, is as follows:

Solution A:

Rochelle salt	170 g/L
Sodium hydroxide	50 g/L
Copper sulfate	35 g/L
Sodium carbonate	30 g/L
Versene-T	20 g/L

Solution B:

Formaldehyde	37% by weight

It is suggested that, immediately prior to use, five volumes of solution "A" be mixed with one volume of solution "B". The solution should be used at room temperature.

Another useful formula reported by the same author is:

Copper sulfate (penta hydrate)..	5 g/L
Sodium hydroxide	7 g/L
Formaldehyde (37% w/v)..........	10 ml/L
Rochelle salts	25 g/L

This is considered a rather stable solution provided it is kept free of nucleating particles such as dust.

Many other copper film formulas are in the literature.

Copper film reduction has become an important process in printed circuitry.

When plating on these mirror films one must consider:

1. Conductivity—the film is thin and, therefore, not too conductive. Too high an initial current density will burn the conductive film. Always start with a low current density and increase it to the desired level as the metal film builds up.

2. Temperature—plating baths should be operated as is reasonably consistent with the temperature stability and expansion characteristic of the plastic. A good rule of thumb is not to use a temperature in excess of 95°F for thermoplastics (cellulose acetate, styrene, methyl methacrylate, etc.) and 130°F for thermosets (ureas, phenolics, etc.)

3. Stability—(of the plastic and mirror film in the plating bath.)

The metal film is quite thin and may be reacted upon before plating begins. Chromium and gold cannot be plated directly on the mirror film. It is best to start plating on a copper film from a nickel bath and on a silver film from an acid copper bath. For silver, cyanide silver baths have also been found satisfactory as initial film thickeners. This is followed with a normal copper electrodeposition of 0.001" to 0.005", polishing and finish plating such as with brass, nickel, gold, etc.

In the last two years there has been much published information and revived interest using the mirror process as the initial conductive film in the application of thick metal films to plastics.

Much attention has been given to acrylonitrile-butadiene-styrene (ABS) as a plastic base for metallizing. Along with this revived interest have come proprietary chemicals claiming excellent results. However, the key to success of any process for electroplating on a non-conductor still involves very careful cleaning, the removal of all cleaning compounds by proper neutralization, and adequate rinsing, surface conditioning, and sensitizing, always using as much rinse water as possible between each step.

Along these same lines there has been recent literature published on the application of metal films to fluorocarbon resins. Here again, the procedure follows along the same lines; however, the initial preparation calls for the conditioning of the fluorocarbon resin in a solution composed of anhydrous ammonia and an alkali metal such as sodium or in a complex prepared by mixing the metal sodium with naphthalene in the solvent tetrahydrofuran. The latter method is considered more practical. The actual solution is made by first mixing one mole or 120 grams of naphthalene in one liter of tetrahydrofuran and then adding one mole or 23 grams of sodium metal (cut up into small pieces). This is all stirred together for two hours at room temperature. Air should be excluded as much as possible during preparation, use and storage. This mix deteriorates with time; therefore, make sure it is still effective if it is stored for any length of time. Following this treatment, the normal steps of sensitizing, nucleating, and then mirroring with a copper film prior to electroplating are carried out.

In some applications, the production of a heavy metal film is not sought; the original properties of the plastic are perfectly adequate for the application.

The only purpose is to produce a film whose appearance might be enhanced if it imitates that of a solid metal object. The silver mirror film and the film produced by the vacuum plating of aluminum are the most common thin films used for this imitative appearance. There is a special surface lacquering before and after the application of the metal film. This before-and-after surface treatment is the same as for vacuum metallizing and will be discussed with that subject. The "Brashear" formula may be used, but one must be careful to remove the work before too heavy a film forms, which loses its luster.

For the most part, this type of film is best produced by spraying. The sensitizing is done with a single nozzle gun spraying sensitizer (stannous chloride solution), water washing then spraying the silver film by use of a double nozzle spray gun. This is a rather unique gun which proportions out a spray of silver fulminate from one nozzle and a spray of reducing agent from the other. These sprays are made to meet and mix at a point about six to eight inches from the nozzle heads. The work to be coated is kept in the mixing plane until a satisfactory mirror forms. This usually takes a few seconds. The solutions used are most often proprietary mixes consisting of a silver fulminate solution made by just redissolving the silver precipitate that first forms by the addition of ammonia (approximately 10 g of silver nitrate and 10 cc of ammonium hydroxide per liter of water), and a reducer made of a hydrazine salt with some alkali.

A typical spray silvering solution and reducer consists of the following:

Silver Solution:

    Silver nitrate .... 2.5 oz/gal
    Ammonia .... 60 ml/gal approx.

Reducer Solution:

    37% Formaldehyde .... 270 cc/gal
    Triethanolamine .... 25 cc/gal

Ammonia is added to the silver nitrate solution until the precipitate which forms just dissolves. Note: Concentrated ammonia and silver nitrate may form explosive mixtures. Therefore, both should be diluted with water before mixing.

All spray operations are carried out in a spray booth made of stainless steel.

Catalytic Deposition

Nickel:

Another method of applying a metal film on non-conductors is by catalytic deposition of nickel, sometimes referred to as "Electroless Nickel." This is possible with thermoset type plastics. Some thermoplastics can also be coated, but only with caution and care because of the temperature-instability characteristic of this group of plastics.

A.S.T.M. publication No. 265 suggests the following preparatory procedure:

1. Clean

2. Roughen

3. Sensitize in stannous chloride at 80°F (70 g/L stannous chloride and 40 g/L hydrochloric acid)

4. Rinse

5. Immerse in cold palladium chloride solution (1 g/L), containing 1 ml of concentrated hydrochloric acid

6. Rinse

7. Immerse at 200°F in the following solution:

Nickel chloride	30 grams
Sodium hypophosphite	10 "
Sodium citrate	10 "
Water	1000 "
pH	4 to 6
Plating rate	0.2 mil per hr

The literature now shows many adaptations of the hypophosphite reduction procedure which may be applied to cobalt, nickel-cobalt alloys, iron-nickel alloys, etc.

Aluminum:

A procedure for the catalytic deposition of aluminum has been described in U.S. Patent 3,462,288. A typical example is to prepare an aluminum hydride solution in a moisture-free nitrogen atmosphere by mixing 49 ml of 1.0 molar lithium aluminum hydride, 18.5 ml of 0.98 molar aluminum chloride and 156 ml of diethyl ether. After stirring, the solution is decanted to produce 0.3 molar aluminum hydride in diethyl ether.

The substrate to be coated is first immersed in a 0.046 molar solution of titanium tetrachloride in diethyl ether, dried at 100°C, cooled to room temperature and immersed in the aluminum hydride solution and again dried. The aluminum film develops in a few minutes. (All these steps should be done in a moisture-free, nitrogen atmosphere.) The appearance of the aluminum film will be a reflection of the substrate: lustrous on a smooth shiny surface, dull on a mat surface.

The future may involve the use of organic coatings which are electrically conductive or plastics themselves which are sufficiently conductive so as to be capable of direct plating without the prior application of some metal film. There are already in experimental existence plastics with much greater conductivity than normal. However, the greatest conductivity achieved is said to be of the order of 10^{-3} reciprocal ohm-cm. This puts such materials somewhere between a plastic insulator having a conductivity of 10^{-10} reciprocal ohm-cm and a good conductor, such as copper, with a conductivity of 10^{5} reciprocal ohm-cm.

Section 10

Useful Literature Sources

Books

"Hess's Paint Film Defects. Their Causes and Cure," Edited and Revised by H. R. Hamburg and W. M. Morgans, 3rd Edition, Chapman and Hall, 504 pp. (1979).

"Plastics vs Corrosives," R. B. Seymour, John Wiley and Sons, 285 pp. (1982).

"Photodegradation and Photostabilization of Coatings," Edited by S. P. Pappas and F. H. Winslow, ACS Symp. Series 151, 308 pp. (1981).

"Federation Series on Coatings Technology," a series of 27 booklets, Federation of Societies for Coatings Technology, Philadelphia, Pa.

> Note: These booklets were published and revised at different times. They may be obtained collectively from the Federated Societies of Coatings Technology.

"Technology of Paints, Varnishes and Lacquers," C. R. Martens, Editor, Reinhold, 744 pp. (1968).

Journals

Progress in Organic Coatings, published bimonthly by Pergamon Press.

Journal of Coatings Technology, published monthly by Federation of Societies for Coatings Technology.

Section 11

Elements of Polymer Science of Importance to Coatings

Polymers are, as the name implies, made up of repeat units of a monomer. Polymers of interest in coatings science have molecular weights of the order of thousands to millions corresponding to repeat units of the order of hundreds to ten-thousands. In the simplest cases, the structural form of a polymer may be considered as a linear polymer or as a polymer network as schematically shown in Figure 11-1. In the practical sense, polymers are generally coiled and the conceptualization in Figure 11-1 represents an idealized form.

The two major types of polymers are (a) addition or chain polymers in which carbon is the only element in the backbone, and (b) condensation or step polymers in which atoms other than carbon are present in the chain or backbone of the polymer. Polyacrylonitrile is an example of an addition polymer:

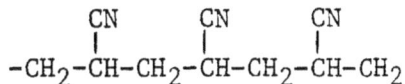

$$\begin{array}{ccc} CN & CN & CN \\ | & | & | \end{array}$$
$$-CH_2-CH-CH_2-CH-CH_2-CH-CH_2$$

backbone: C-C-C-C-C-C-C-

Nylon-6 is an example of a condensation polymer:

$$-NH-(CH_2)_5-CO-NH-(CH_2)_5-CO-NH-(CH_2)_5-CO-$$

backbone: N-C-C-C-C-C-C-N-C-C-C-C-C-C-N-C-C-C-C-C-C-.

Some examples of the building blocks for addition polymers are given in Table 11-1 and some examples of the building blocks for condensation polymers are given in Table 11-2.

Polymerization Process

The polymerization process may be carried out in many different ways. The process selected depends upon the system and the planned application since the properties of the resulting product are a function of the method of polymerization used.

unbranched   branched

Linear Polymers

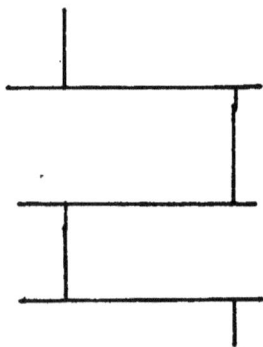   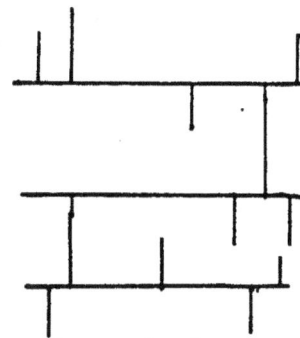

unbranched   branched

Polymer Networks

Figure 11-1. Schematic form of different types of stretched
polymers. Picture is idealized since polymers
are normally coiled.

Vinyl chloride	$CH_2=CHCl$
Ethylene	$CH_2=CH_2$
Vinyl alcohol	$CH_2=CHOH$
Propylene	$CH_2=CH-CH_3$
Vinyl acetate	$CH_3-COO-CH=CH_2$
Butadiene (1,3)	$CH_2=CH-CH=CH_2$
Styrene	$C_6H_5CH=CH_2$
Acrylonitrile	$CH_2=CH-CN$
Acrylic acid	$CH_2=CH-COOH$
Ethyl acrylate	$CH_2=CH-COOC_2H_5$
Methyl methacrylate	$CH_2=C(CH_3)-COOCH_3$
Acrylamide	$CH_2=CH-CONH_2$

Table 11-2

Building Blocks for Condensation Polymers

Polyester	$-\overset{\overset{O}{\parallel}}{C}-O-$
Polyamide	$-\overset{\overset{O}{\parallel}}{C}-NH-$
Polyanhydride	$-\overset{\overset{O}{\parallel}}{C}-O-\overset{\overset{O}{\parallel}}{C}-$
Polyacetal	$-O-\overset{\overset{H}{\mid}}{\underset{\underset{R}{\mid}}{C}}-O-$
Polyurea	$-NH-\overset{\overset{O}{\parallel}}{C}-NH-$
Polyurethane	$-O-\overset{\overset{O}{\parallel}}{C}-NH-$
Polysiloxane	$-\overset{\mid}{\underset{\mid}{Si}}-O-\overset{\mid}{\underset{\mid}{Si}}-$
Polysulfide	$-S-S-$

Four major methods are used in the polymerization of addition polymers: bulk polymerization, solution polymerization, suspension polymerization, and emulsion polymerization. Bulk polymerization is exemplified by polystyrene which can be polymerized at intermediate temperatures of the order of 60°C by the addition of an initiator such as benzyl peroxide or stannic chloride to the undiluted monomer in liquid form. The quality of the resulting product is largely a function of the speed with which the reaction occurs since it is highly exothermic. High quality castings may be obtained when the reaction rate is low and frothy, cracked or void-filled castings may be obtained if the reaction rate is too rapid. Bulk polymerization is rarely used commercially because of the difficulty of obtaining a high quality product and because the product requires additional forming steps.

Some of the disadvantages of bulk polymerization are removed by diluting the reactants in an inert solvent. If both the starting material and the final product are soluble, it is possible to control the reaction and the heat generated by stirring the solution. Practical difficulties in the process include restrictions in the choice of solvent, the high viscosity that often results as polymerization proceeds, and the inability in many cases to rid the polymer completely of the solvent.

Suspension polymerization is carried out by dispersing the monomer in a liquid in which the monomer is insoluble. Polymerization occurs in isolated droplets whose size may be controlled in the range of 0.1 to 3 mm. The resulting product is granular in nature and the shape lends itself to normal industrial processing. Problems encountered with suspension polymerization may occur during the intermediate stage of polymerization when the droplets are not fully polymerized and are "sticky" in nature. The particles may clump and may adhere to the walls of the vessel and to the stirring equipment. Such problems are usually prevented by the addition of additives to the solution that change the surface character of the suspended polymer particles.

Emulsion polymerization is a special type of suspension polymerization in which the dispersed monomer particles are much smaller in size. The initiator is generally added to the solvent phase, rather than the monomer phase but hybrid processes are used in which the initiator is soluble in the monomer as well. Emulsion polymerization has achieved special recognition because the process lends itself to the preparation of polymer particles of uniform size, so-called monodisperse polymers.

The overall polymerization process is divided for convenience in mathematical treatment into three stages: initiation, propagation, and termination. The initiation stage depends upon the presence of an active molecule which activates the double bonds in the monomer so that the bonds are rearranged and a larger molecule results. The propagation stage is the continuation of addition of monomer units to the growing large molecule. Termination occurs when two activated molecules react, thus removing these two molecules as active constituents.

Procedures for preparing condensation polymers vary widely depending upon the system and the desired end use of the product. Modifications and elaborations of the same procedures used with addition polymers are common.

Crystalline Nature of Polymers

The properties of polymers depend not only on the size, shape, and chain structure of the individual units, but also on the spatial shape of the polymer molecule. The linking of many carbon atoms and the freedom of rotation about carbon-carbon bonds permit the molecule to assume a variety of spatial shapes such as spirals, coils, and tangles. This wide latitude in shape also leads to a variety of ways in which individual molecules are oriented with respect to their neighbors. Three classes of arrangement are recognized:

(1) Segments of the molecule are randomly distributed regardless of whether they belong to the same molecular chain or another chain. Such a material is termed amorphous or glassy and the properties are uniform in all directions.

(2) Segments of the molecule possess a degree of lateral order through the folding of individual chains. The volume element over which this occurs may be considered a single crystal. The individual crystals may be randomly oriented or they may be aligned in the same direction. In the latter case the polymer may have physical properties that differ in different directions.

(3) Segments of the molecule may show lateral order through the parallel arrangement of extended chains. As in (2), these parallel arrangements may be unoriented with respect to neighboring volumes or there may be a degree of spatial orientation. Materials of this type are obtained when a polymer melt is solidified while under shear or stress.

The degree of orientation, or crystallinity, is determined by diffraction techniques using either X-rays or electrons. Highly oriented products are obtained when polymeric materials are formed by spinning with subsequent drawing or rolling. Processes such as extrusion, injection or comparison lead to products with lower degrees of crystallinity.

The Glass Transition Temperature of Polymers

The glass transition temperature of a polymer, T_g, is that temperature at which a discontinuity occurs in a physical property as a function of temperature when the polymer exists in the amorphous (non-crystalline) condition. Typical physical properties which show discontinuities at T_g include the coefficient of expansion and specific heat. It is interpreted as that temperature above which the polymer has sufficient thermal energy for isomeric rotational motion or for significant torsional oscillation to occur about most of the bonds in the main chain which are capable of such motion. Values of T_g are obtained by many different experimental techniques, the more common of which include dilatometry, dielectric measurements, spectroscopy, calorimetry, and refractive index. The T_g of polymers used in coatings is important because physical properties such as gas permeability and ductility differ above and below T_g.

Mechanical Properties of Polymers

The mechanical properties of polymers depend on molecular weight, crystallinity and the three-dimensional arrangement of branches. An increase in molecular weight makes a polymer harder and stronger. The higher the degree of crystallinity, the stronger the polymer. Chain polymers containing two different groups, R and R', have different mechanical properties dependent upon the arrangement of the branches. Three main types are recognized:

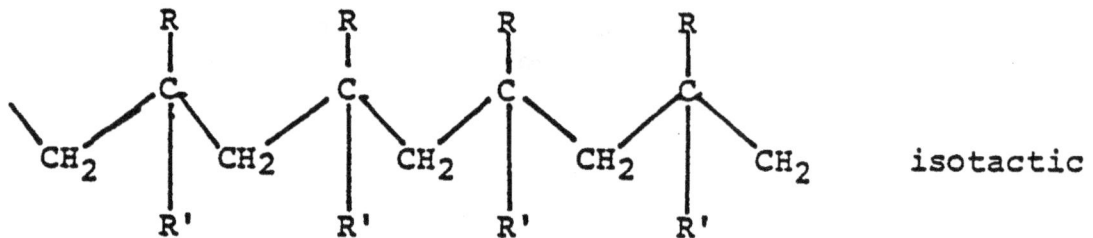

isotactic

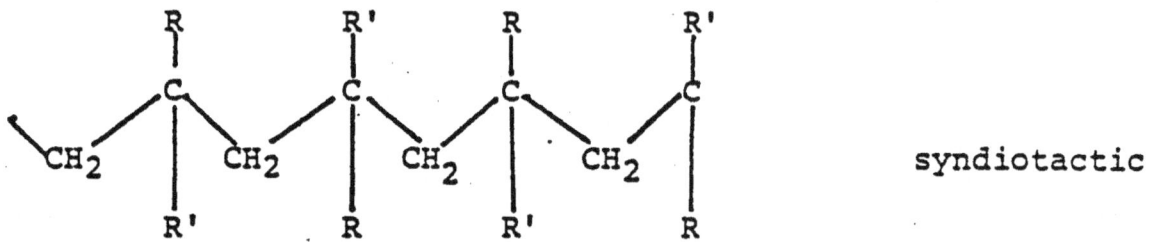

syndiotactic

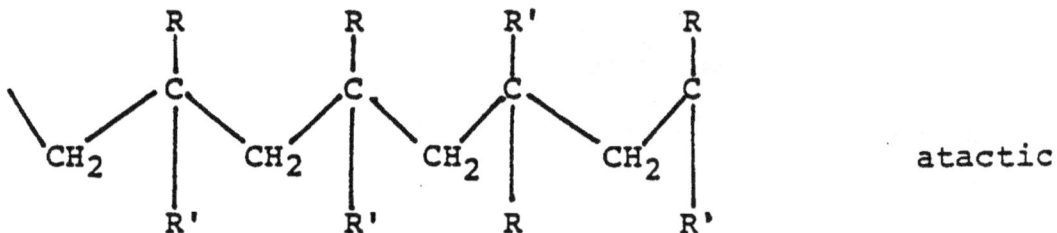

atactic

Curing or cross-linking causes a polymer to become harder, more brittle, and less soluble.

Two classes of polymers are distinguished on the basis of their reaction to heating. Thermoplastic polymers soften at higher temperature but re-harden on cooling. Thermosetting polymers become harder on heating, through further cross-linking, and do not regain their original properties on cooling.

Environmental Degradation of Polymers

Polymer coatings are exposed to the environment and thus are subject to degradation by environmental constituents. The main agencies by which degradation occurs are thermal, mechanical, radiant and chemical. Deterioration takes the form of discoloration, cracking or crazing, loss of adherence to the substrate, or change in a physical property such as resistivity or mechanical strength. The mode of degradation may involve depolymerization, generally caused by heating, splitting out of constituents in the polymer, chain scission, cross linking, oxidation, and hydrolysis. Polymers may also be degraded by living organisms such as mildew. Polymers are subject to cracking on the application of a tensile force, particularly when exposed to certain liquid environments. This phenomenon is known as environmental stress cracking or stress corrosion cracking.

Use of Polymers in Coatings

Polymers form the matrix of organic coatings. They are the retainers for pigments, fillers, and other additives present for specific purposes. The polymer selected is based both on end use requirements and the ability to apply the coating in the desired manner. Emulsion polymers, or latexes, are suspended in an aqueous medium and they form a coating by loss of water by evaporation and coalescence of the individual particles into a continuous film. Some monomers, such as butadiene, are applied to a substrate in a solvent and the polymerization process occurs as the solvent is removed. Cross-linking by oxidation occurs in the case of polybutadiene. Other coatings based on condensation polymers are polymerized *in situ*. A good example is the epoxy-polyamine coating in which the two constituents are mixed just prior to application and the polymerization process occurs over a period of time after application. Other polymers are dissolved in solvents and the polymer forms the coating as the solvent is evaporated. Production line painting often involves the application of heat or other type of radiation to cause the film-forming process to occur more rapidly.

Section 12

How Paint Films are Formed

The paint user has only temporary interest in the liquid paint. His long term concern is the dried film. Different kinds of paint dry in radically different ways and the nature of the process of drying or film formation profoundly affects film properties. The broadest classification of paints with regard to the process of film formation is as thermoplastic and conversion. Thermoplastic coatings dry solely by evaporation of the solvent, without chemical or physical change of the non-volatile material that forms the film. On the other hand, the drying of conversion coatings involves first solvent evaporation and then a change of the liquid or soft residue to a hard, usable film. The process of conversion may be affected in different ways, so there are several types of conversion coatings.

Shellac varnish is a good example of a thermoplastic coating. It is made by simply dissolving gum shellac in alcohol. When the varnish is applied, the alcohol evaporates, leaving on the surface a thin film of unchanged shellac resin. A conventional nitrocellulose lacquer dries in the same way. Naturally films formed in this manner are readily attacked and softened or removed by the original solvent or similar solvents. Moreover, if the resin is heated below the decomposition temperature it softens and, on recooling, it hardens again. Consequently, the term thermoplastic.

All conversion coatings undergo a chemical and physical change in the process of film formation. There are several types of conversion coatings, according to the manner of conversion. Paints that have a drying oil or an oleoresinous varnish or resin as the binder usually dry more slowly than thermoplastic coatings and in overlapping stages. The stages are solvent evaporation, oxidation, thickening or polymerization, and gellation. Gellation occurs when the polymers reach a size and concentration such that they form a continuous network. Although the film is now considered dry, it contains much liquid material and is soft. Conversion of the remaining liquid material to solid material continues gradually until the coating becomes hard and finally brittle. These changes are greatly accelerated by heat and sunlight. This process of film formation is known as oxygen conversion and the coatings are classified as oxygen convertible. Generally speaking the films are flexible, somewhat soft and lack high resistance to water and chemicals.

Typical baked coatings present another important method of film formation, by heat conversion. In the liquid state heat convertible coatings are solutions of one or more resin polymers. Under the influence of heat the polymers increase in size and join to form a solid film. When the polymers are of two or more chemical types, there may be a definite cross-linking reaction. The most important examples of heat convertible resins are urea-formaldehyde, thermosetting acrylic, and mixtures of epoxy and urea-formaldehyde. The fact that a paint is baked does not necessarily mean that the binder is heat convertible.

An oxidizing alkyd may be baked to accelerate oxygen conversion. Some thermoplastic binders, e.g. vinyl copolymers and thermoplastic acrylics, are ordinarily baked to improve levelling, gloss, and adhesion. As a class, heat convertible coatings are hard and resistant to solvents and chemicals. They have a wide range in degree of flexibility.

During recent years another method for film formation has become important, by catalyst conversion or by cross-linkage, without the necessity of heating. The most common example of true catalyst conversion is the addition to a urea-formaldehyde resin of an acid such as n-butyl phosphoric acid or p-toluene sulfonic acid. Cross-linkage is exemplified by mixtures of an epoxy resin with an amine. In the latter case the amine becomes an integral part of the new polymer, while in the former case the acid does not. The two mechanisms of conversion have in common the feature of limited pot life, which restricts or prevents application by dipping or flowcoating.

When the binder in a water emulsion paint is vinyl resin or acrylic resin, the film is thermoplastic type. If, however, the binder is a polymerized drying oil or an oxidizing alkyd resin, the film is oxygen convertible. In either case the initial stage of film formation is the same: by fusion or coalescence of the suspended resin particles when they become crowded closely together by evaporation of the water. With thermoplastic resins film formation is completed with fusion. If the resin is oxygen convertible, fusion is followed by oxygen conversion. Since fusion is likely to be incomplete, especially in highly pigmented products, emulsion films tend to be more permeable to water and moisture than films deposited from solutions. Vinyl organosols and vinyl plastisols represent modifications of film formation by coalescence.

Although there are other methods of film formation, they have secondary importance and do not warrant attention here. It should be mentioned, however, that frequently a paint contains film formers of more than one type. A common example is nitrocellulose lacquer containing oxygen convertible alkyd resin or oil. Then there are baking enamels in which the heat convertible urea or melamine resin is combined with oxygen convertible alkyd and/or thermoplastic alkyd.

Thus there is another concept which may be applied to paint. It is that the liquid (or solid) paint is really not a finished product: that the last step in manufacturing the finished product (the cured film) must be taken by the user, beyond the direct control of the manufacturer, and that the "quality" of the product, that is its performing the job intended for it, depends on the success of the manufacturer in influencing the user to select the right paint for the job and to apply it properly to a correctly prepared surface.

Table 12-1

Approximate Film Thickness of Paint Systems

System	Mils
2 Coats of oil base house paint	3
2 Coats of emulsion house paint	1.5 – 2.0
3 Coats of oil or varnish base exterior metal paint	3 – 4
Factory painted aluminum siding	1
2 Coats of floor varnish	1.5
2 Coats of Spar varnish	2
Kitchen appliances	2
Automobiles	2.0 – 2.5

Section 13

Common Organic Coatings and Their Properties

NOTE: The information contained in the following table is taken from
the book, "Design and Corrosion Control", V. Roger Pludek,
Halsted Press, 1977, 383 pp.

	Alkyd	Alkyd amine	Alkyd phenolic	Alkyd silicone	Alkyd urea	Styrenated alkyd	Acrylic	Bituminous
Physical properties								
Sward rocker hard (8th day)	24	30	34	16–30	28	28	24	—
Flexibility	E	VG	G	VG	VG	G	E	E
Abrasion res. cyc. (Taber)	3.5K	>5K	>5K	4K	>5K	>5K	2.5K	—
Max svc temperature (°C)	93	121	121	232	107	93	82	93
Toxicity	none	slight	none	none	slight	slight	none	—
Impact resistance	VG	E	G	G	E	G	E	E
Dielectric properties	G	G	VG	E	G	G	VG	—
Adhesion to:								
ferrous metals	E	E	E	VG	E	F	VG	E
non-ferrous metals	F	E	E	VG	VG	F	VG	E
old paints	VG	G	G	E	G	VG	P	—
Decorative properties								
Colour retention	G	VG	P	E	VG	G	E	—
Initial gloss	E	E	VG	E	E	E	E	P
Gloss retention	E	G	F	E	F	G	E	—
Chemical resistance								
Atmospheric, exterior	E	E	E	E	E	F	E	F
Salt spray	E	VG	E	E	G	G	E	E
Solvents, alcohols	F	G	G	G	G	G	P	P
Solvents, gasoline	G	E	E	E	E	F	G	P
Solvents, hydrocarbons	G	E	E	G	E	E	F	P
Ammonia	P	P	P	P	P	P	P	—
Alkalis	P–F	G–VG	P–F	G–VG	G	G–VG	F–G	E
Acids, oxidising	P	P–F	P–G	P	P–F	P–F	P–F	—
Water (salt, fresh)	F	G	G	G	F	G	E	E
Application								
Ease of application	E	bake	E	E	bake	E	VG	P–VG
Priming required	PR	none	none	PR	none	none	PR	none
Solvent for application	HyC	HyC	HyC	HyC	HyC	HyC	blend	—
Method	U	U	U	U	U	U	U	U
Cure	A or B	B	A or B	A or B	B	A or B	A	A
Baking temperature (°C)	135	160	177	177	160	149	--	—
Bake drying time	30 min	20 min	30 min	30 min	20 min	15 min	—	—
Air drying times:								
touch	2H	—	20 min	45 min	—	10 min	5 min	2H
handle	4H	—	60 min	2H	—	30 min	15 min	24H
re-coat	4H	—	6H	4–6H	—	4H	15 min	—
hard	12H	—	6H	12H	—	4H	12H	24H
corrosion resistant	48H	—	48H	12H	—	48H	24H	—
Coverage (ft²/gal/mil; m²/l/μm)	450; 280	450; 280	450; 280	500; 312	450; 280	400; 249	350; 221	—
Average dry film thick. mil; μm	1.5; 38	1.5; 38	1.5; 38	0.6; 15	1.5; 38	1.5; 38	1.0; 25	3–250; (75–6350)
Cost	L						M	

*A, air dried; B, baked; E, excellent; F, fair; G, good; H, high; L, limited or low; M, medium; MH, medium high; P, poor; S, slightly limited; U, unlimited; VG, very good.

	Epoxy amine	Epoxy ester	Epoxy furane	Epoxy melamine	Epoxy phenolic	Epoxy urea	Furane	Phenolic
Physical properties								
Sward rocker hard (8th day)	36	30	24	36	44	34	38	38
Flexibility	F	E	E	VG	VG–E	VG	F	G
Abrasion res. cyc. (Taber)	>5K	>5K	—	>5K	>5K	>5K	—	>5K
Max svc temperature (°C)	204	149	177	204	204	204	149	177
Toxicity	none	none	none	none	none	none	none	none
Impact resistance	G	E	G	VG	VG	G	F	G
Dielectric properties	VG	VG	G	VG	VG	VG	F	E
Adhesion to:								
ferrous metals	E	E	E	E	E	E	F	E
non-ferrous metals	E	E	E	E	E	E	F	E
old paints	G	VG	E	P	P	P	E	G
Decorative properties								
Colour retention	F	G	G	G	P	G	G	P
Initial gloss	E	E	G	VG	VG	VG	E	VG
Gloss retention	F	G	G	F	F	F	F	F
Chemical resistance								
Atmospheric, exterior	G	E	E	E	E	VG	G	E
Salt spray	VG	E	E	E	E	E	G	E
Solvents, alcohols	G	F	E	E	E	E	E	E
Solvents, gasoline	E	E	E	E	E	E	E	E
Solvents, hydrocarbons	E	VG	E	E	E.	E	E	E
Ammonia	G	P	E	P	F	P	E	P
Alkalis	E	E	E	E	E	E	E	P
Acids, oxidising	P–G	P–F	F	P–G	P–E	P–F	P	P–F
Water (salt, fresh)	G	VG	E	G	E	G	E	E
Application								
Ease of application	cat	E	E	bake	bake	bake	E	E
Priming required	none	PR	none	none	none	none	PR	none
Solvent for application	blend	HyC	ket	blend	blend	blend	ket	alc
Method	L	U	U	L	L	L	U	U
Cure	A	A or B	B	B	B	B	A or B	A or B
Baking temperature (°C)	—	160	177	177	204	177	149	177
Bake drying time	—	30 min	30 min	30 min	30 min	30 min	30 min	30 min
Air drying times:								
touch	60 min	1H	—	—	—	—	1H	10 min
handle	2H	2H	—	—	—	—	4H	30 min
re-coat	6–8H	8H	—	—	—	—	6H	30 min
hard	12H	8H	—	—	—	—	24H	4H
corrosion resistant	7–10D	5D	—	—	—	—	48H	24H
Coverage (ft²/gal/mil; m²/l/μm)	500; 312	450; 280	450; 280	500; 312	450; 280	500; 312	400; 249	350; 221
Average dry film thick. (mil; μm)	1.8; 45	1.5; 38	0.5–1.0; 12–25	1.8; 45	1.8; 45	1.8; 45	3–5; 76–127	1.5; 38
Cost	H							

	Polyamide (nylon)	Polyester	Silicone	Polyethylene	Chlorinated rubber	Neoprene rubber	Hypalone rubber	Viton
Physical properties								
Sward rocker hard (8th day)	—	30	16	F	24	<10	<10	<10
Flexibility	G	G	F	E	VG	E	E	G
Abrasion res. cyc. (Taber)	—	3.5K	2.5K	—	>5K	5K	5K	1K
Max svc temperature (°C)	149	93	288	93	93	93	121	288
Toxicity	—	none	none	none	slight	none	—	slight
Impact resistance	VG	F	F	F	G	E	E	E
Dielectric properties	G	G	E	E	E	G	VG	G
Adhesion to:								
ferrous metals	VG	F	F	E	F	VG	VG	VG
non-ferrous metals	VG	P-F	E	E	VG	VG	VG	VG
old paints	—	P	E	—	—	—	—	—
Decorative properties								
Colour retention	—	G	E	VG	G	G	E	G
Initial gloss	G	G	E	VG	F	P	P	E
Gloss retention	—	F	E	VG	F	F	F	F
Chemical resistance								
Atmospheric, exterior	P	VG	E	P	E	E	E	E
Salt spray	F	G	E	VG	E	E	E	E
Solvents, alcohols	G	G	F	E	E	E	—	E
Solvents, gasoline	G	E	F	P	G	G	G	E
Solvents, hydrocarbons	—	G	VG	VG	—	—	—	—
Ammonia	G	P	P	E	G	G	G	E
Alkalis	G	P	F-E	P-G	E	E	E	E
Acids, oxidising	—	P	P	VG	F-E	F-P	F-G	E
Water (salt, fresh)	F	G	E	VG	E	E	E	G
Application								
Ease of application	G	F	E	E	G	VG	VG	VG
Priming required	none	PR	PR	none	PR	none	PR	PR
Solvent for application	—	styr	HyC	—	HyC	HyC	HyC	blend
Method	L	L	U	SL	U	U	U	L
Cure	A	A or B	A or B	B	A or B	A or B	A or B	A or B
Baking temperature (°C)	—	149	232	232	149	149	149	—
Bake drying time	—	15 min	1H	15 min	15 min	15 min	15 min	—
Air drying times:								
Touch	—	1H	45 min	—	45 min	15 min	15 min	5 min
handle	—	1H	2H	—	2H	30 min	30 min	15 min
re-coat	—	1H	4-6H	—	4-6H	4H	4H	12H
hard	—	12H	12H	—	4-6H	4H	4H	12H
corrosion resistant	—	7-10D	12H	—	24H	7-10D	7-10D	12H
Coverage (ft²/gal/mil; m²/l/μm)	—	800;	350;	560;	450;	300;	250;	200;
	—	498	221	348	280	186	155	125
Average dry film thick. (mil; μm)	2-30;	2.0;	1.0;	3-10;	1.5;	2-10;	2.0;	1.0;
	50-762	50	25	76-254	38	50-254	50	25
Cost		M						

	Urethane	Vinyl	Vinyl alkyd (1:1 approx)	Vinyl organosol	Vinyl plastisol	Nylon powder	Cellulosic	Chlorinated polyether
Physical properties								
Sward rocker hard (8th day)	35–65	20	26					
Flexibility	E	E	E	E	E	G	G	F
Abrasion res. cyc. (Taber)	>5K	>5K	2.5K					
Max svc temperature (°C)	149	66	82		93	82	82	121
Toxicity	slight	none	none					
Impact resistance	E	E	E	E	E	VG	E	G
Dielectric properties	E	E	G		VG	G	VG	VG
Adhesion to:								
ferrous metals	E	G	VG	E	E			
non-ferrous metals	E	VG	G					
old paints	—	—	—					
Decorative properties								
Colour retention	G	VG	E	G	VG	VG	E	G
Initial gloss	E	G	E	P	VG	G	E	G
Gloss retention	F	E	E		G	—	E	—
Chemical resistance								
Atmospheric, exterior	E	E	E	E	E	F	E	F
Salt spray	E	E	E	E	E	G	E	E
Solvents, alcohols	VG	F	G	F	E	G	F	E
Solvents, gasoline	F–G	E	E	F	E	E	G	VG
Solvents, hydrocarbons	—	—	—	—	G	E	G	E
Ammonia	P	E	P	P	E	G	P	E
Alkalis	F–VG	E	P–G	F	E	G	F	E
Acids, oxidising	P–G	G–E	P	P	E	P	P	E
Water (salt, fresh)	E	E	E	E	E	F	VG	E
Application								
Ease of application	E	F	VG	G	G			
Priming required	PR	PR	PR					
Solvent for application	blend	blend	blend	blend	none			
Method	U	L	U	L	L	L	L	L
Cure	A or B	A or B	A					
Baking temperature (°C)	163	149	—					
Bake drying time	30 min	15 min						
Air drying times:								
touch	45 min	15 min	5 min					
handle	1–2H	30 min	15 min					
re-coat	4–6H	4–6H	15 min					
hard	18H	4–6H	12H					
corrosion resistant	5–7D	24H	24H					
Coverage (ft²/gal/mil; m²/l/μm)	—	250; 155	200; 125					
Average dry film thick. (mil; μm)	1.4; 36	1.0; 25	1.0; 25					
Cost		MH	M	L	L			

Section 14

Constituents Commonly Present in Coatings and Their Purpose

Liquid Carrier. The carrier, or solvent as it is often called, may be an organic material or in the case of latexes it may be water.

The compatibility of the solvents with resins is measured in terms of the solubility parameter, which is defined as:

$$S = (E_v/V_1)^{\frac{1}{2}}$$

where S is the solubility parameter, E_v is the isothermal energy of vaporization at zero pressure in calories per mole and V_1 is the molar volume in cc per mole. The term (E_v/V_1) represents the mutual attraction of the compound's molecules and is called the cohesive energy density. Values of the solubility parameter for (a) poorly hydrogen-bonded solvents, (b) moderately hydrogen bonded solvents, (C) highly hydrogen-bonded solvents, and for polymers and plasticizers are given in the accompanying 4 tables (Table 14-1-4).

The evaporation rate of the solvent while the coating is drying is very important. Representative values for several organic liquids are given in Table 14-5 and the effect of temperature is given in Table 14-6.

Pigments. These insoluble materials are added to the coating to give it color and to reduce the rate at which water and gases pass through the coating. A list of some typical white pigments and their properties are given in Table 14-7 and Table 14-8. Some typical colored pigments and their properties are given in Table 14-9.

Binders or (Matrix). The work "binder" refers to the polymer or resin that remains after the solvent evaporates and which holds the other components of the coating. The various matrix materials have been discussed in Sections 11 and 13.

Plasticizers. A plasticizer is a substance incorporated into a polymer to increase its workability, flexibility, toughness and extensibility. When these effects are achieved by chemical modification of the polymer molecule, e.g. through copolymerization, the resin is said to be internally plasticized. Some representative plasticizers are given in Table 14-10.

Drier. A drier is an additive to coatings which accelerates or controls the hardening of applied films by catalyzing the oxidation polymerization of oleoresinous vehicle components. To be effective, driers must be miscible in unpolymerized and polymerized drying oils and resins, in mixtures of these and in solvent thinners normally used in coating formulations.

Flow Modifiers. Additives are often used to change the rheological properties of the liquid coating. Metal soaps are often used.

Table 14-1

Solubility Parameter of Solvents (Poorly Hydrogen Bonded)

Group I Solvents	Solubility Parameter	Group I Solvents	Solubility Parameter
Silicones	5.5	p-Xylene	8.7
2,2-Dimethybutane	6.7	p-Diethylbenzene	8.7
Mineral spirits	6.9	m-Diethylbenzene	8.7
Isooctane	6.9	m-Xylene	8.8
n-Hexane	7.3	Mesitylene	8.8
n-Heptane	7.4	Ethylbenzene	8.8
"Freon" 113	7.4	Ethylchloride	8.8
V.M.&P. naphtha	7.6	Toluene	8.9
n-Nonane	7.7	o-Diethylbenzene	8.9
n-Decane	7.7	o-Xylene	9.0
Methylcyclohexane	7.8	Benzene	9.2
n-Dodecane	7.8	Styrene	9.3
n-Tetradecane	7.9	Chloroform	9.3
n-Hexadecane	8.1	Tetrachloroethylene	9.4
Turpentine	8.1	"Tetralin"	9.5
Cyclohexane	8.2	1,1,2-Trichloroethane	9.6
Amyl chloride	8.4	n-Chlorotoluene	9.7
Benzonitrile	8.4	Nitrobenzene	10.0
Dodecylbenzene	8.5	Carbon disulfide	10.0
Propylbenzene	8.6	Bromobenzene	10.3
Pine oil	8.6	1-Nitropropane	10.7
Decalin	8.6	Nitroethane	11.1
Carbon tetrachloride	8.6	Acetonitrile	11.9
Butylbenzene	8.6	Nitromethane	12.7

Table 14-2

Solubility Parameters of Solvents (Moderately Hydrogen Bonded)

Group II Solvents	Solubility Parameter	Group II Solvents	Solubility Parameter
Diisopropyl ether	7.0	Isophorone	9.1
Diethyl ether	7.4	Ethyl Acetate	9.1
Diisobutyl ketone	7.8	Propylene oxide	9.2
Methyl butyrate	8.0	Ethylene glycol monomethyl	
Methylamyl acetate	8.0	ether acetate	9.2
Methyltetrahydrofuran	8.1	Methyl ethyl ketone	9.3
sec-Butyl acetate	8.2	Dibutyl phthalate	9.4
Isobutyl acetate	8.3	Methyl acetate	9.6
sec-Amyl acetate	8.3	Diethylene glycol monoethyl	
Methyl isobutyl ketone	8.4	ether	9.6
Isopropyl acetate	8.4	Benzyl acetate	9.8
Methyl amyl ketone	8.5	Dioxane	9.9
Butyl acetate	8.5	Diethyl ketone	9.9
Amyl formate	8.5	Cyclohexanone	9.9
Amyl acetate	8.5	Ethylene glycol monoethyl ether	9.9
Methyl propyl ketone	8.7	Ethyl lactate	10.0
Ethylene glycol monoethyl		Acetone	10.0
ether acetate	8.7	Methyl formate	10.2
Tetrahydrofuran	8.8	Cyclopentanone	10.4
Propyl acetate	8.8	Acetaldehyde	10.4
Dimethyl ether	8.8	Pyridine	10.7
Methyl isopropyl ketone	8.9	Ethylene glycol monomethyl	
Ethylene glycol monobutyl		ether	10.8
ether	8.9	Ethyl acetamide	12.3
Mesitylene oxide	9.0	Ethyl formamide	13.9
Butyraldehyde	9.0	Methyl acetamide	14.6
Methyl n-butyl ketone	9.1	Ethylene carbonate	14.7
		Methyl formamide	16.1

Table 14-3

Solubility Parameter of Polymers and Plasticizers

Polymer	Solubility Parameter		
	Poorly Hydrogen-bonded Solvents	Moderately Hydrogen-bonded Solvents	Strongly Hydrogen-bonded Solvents
Nitrocellulose	11.9 ± 0.8	11.2 ± 3.4	Insoluble
Ethylcellulose T-10	9.0 ± 0.5	8.8 ± 1.0	10.4 ± 1
45% soya-glycerol-phthalic alkyd	9.1 ± 2.1	9.7 ± 2.3	10.7 ± 1.2
30% soya-glycerol-phathalic alkyd	10.4 ± 1.9	11.6 ± 3.7	Insoluble
Soya oil	9.0 ± 2.0	9.7 ± 2.3	10.7 ± 1.2
45% linseed-glycerol-phthalic alkyd	9.5 ± 2.4	9.7 ± 2.2	10.7 ± 2.2
Polyvinyl chloride-acetate (IYHH)	10.2 ± 0.9	10.6 ± 2.8	Insoluble
Polyvinyl acetate (AYAA)	10.8 ± 1.9	11.6 ± 3.1	Insoluble
Polystyrene KT PL-A	9.3 ± 1.3	9.0 ± 0.9	Insoluble
Ester gum	8.8 ± 1.8	9.1 ± 1.7	10.2 ± 0.7
Phenolic resin, "Durez" 220	9.5 ± 1.1	8.8 ± 1	10.5 ± 1.0
"Epon" 1001	10.8 ± 0.2	11.1 ± 1.1	Insoluble
Chlorinated rubber	9.5 ± 1.0	9.3 ± 1.5	Insoluble

Table 14-4

Solubility Parameter of Solvents
(Highly Hydrogen Bonded)

Group III Solvents	Solubility Parameter
Diethylene glycol	9.1
Diacetone Alcohol	9.2
Ethylhexanol	9.5
Methyisobutyl carbinol	10.0
n-Octyl alcohol	10.3
2-Ethylbutane	10.5
Methoxybutanol	10.6
n-Heptyl alcohol	10.6
n-Hexyl alcohol	10.7
sec-Butyl alcohol	10.8
Amyl alcohol	10.9
Acetic acid	10.9
Isobutyl alcohol (primary)	11.1
Cyclohexanol	11.4
n-Butyl alcohol	11.4
Methylbenzyl alcohol	11.5
Isopropyl alcohol	11.5
n-Propyl alcohol	11.9
Benzyl alcohol	12.1
Furfural alcohol	12.5
Ethyl alcohol	12.7
Ethylene glycol	14.2
Methyl alcohol	14.5
Glycerol	16.5
Water	23.4

Table 14-5

Comparison of Average Evaporation Rates on
Hydrocarbon and Oxygenated Solvents

	Evaporation Rate[a]	Boiling Range	
		(°C)	(°F)
Hydrocarbon Solvents			
Hexanes	5560	66-70	150-158
Fast Diluent naphtha	5290	60-82	140-180
n-Heptane	2860	98-99	208-209
Lacquer diluent	2500	93-116	200-240
n-Octane	1095	125-126	257-258
V.M.&P. naphtha	990	121-149	250-300
Mineral Spirits	80	154-204	310-400
Toluene	1900	110-111	230-232
Xylene	650	135-143	275-290
Oxygenated Solvents			
Isopropyl alcohol	1305	81-83	178-181
n-Butyl alcohol	338	116-119	241-246
Ethyl acetate	4440	72-80	162-176
n-Propyl acetate	2250	95-103	203-217
Amyl acetate	474	120-150	248-302
Ethylene glycol monoethyl ether	330	132-137	270-279
Acetone	5830	56-57	133-135
Methyl ethyl ketone	3620	78-81	172-178
Cyclohexanone	290	130-173	266-343

[a]Grams evaporated per square centimeter per second x 10^8. Measurements were made with the Shell Thin Film Evaporometer.

Table 14-6

Effect of Temperature on Evaporation Rate

| | Evaporation Rate at Indicated Temperature[a] | | | |
	77°F	104°F	140°F	Relative Rate[b]
Hydrocarbon Solvents				
n-Heptane	2860	4960	7440	2.6
n-Octane	1095	2320	4430	4.0
V.M.&P. naphtha	990	--	4350	4.4
Toluene	1900	--	5640	3.0
Xylene	650	1475	2990	4.6
Oxygenated Solvents				
Isopropyl alcohol	1305	2350	3240	2.5
n-Butyl alcohol	338	840	1830	5.4
Ethyl acetate	4440	6310	8500	1.9
n-Propyl acetate	2250	3860	7020	3.1
Amyl acetate	474	--	2630	5.5
Ethylene glycol monoethyl ether	330	760	1840	5.6
Methyl ethyl ketone	290	5560	7450	2.1
Cyclohexanone	290	--	1730	6.0

[a]Grams evaporated per square centimeter per second x 10^8

[b]Evaporation rate at 140°F divided by evaporation rate at 77°F.

Table 14-7

Properties of Supplemental Pigments

Pigment	Approximate Chemical Composition	Refractive Index	Specific Gravity	Bulking Value (gal/100 lb)	Oil Absorption (lbs oil/ 100 lb of pigment)
Barium sulfate pigments					
Ground barytes	$BaSO_4$	1.64	4.476	2.650	6.0
Blanc fixe	$BaSO_4$	1.64	4.35	2.753	14.0
Calcium Carbonate Pigments					
Precipitated types					
Colloidal	$CaCO_3$	1.63	2.68	4.48	55.0
High oil absorption	$CaCO_3$	1.63	2.68	4.48	40.0
Low oil absorption	$CaCO_3$	1.63	2.68	4.48	17.0
Surface treated	$CaCO_3$	1.63	2.65	4.53	15.0
Natural types					
Calcite					
Water ground	$CaCO_3$	1.60	2.71	4.43	15.0 - 6.5
Dry ground	$CaCO_3$	1.60	2.71	4.43	9.0
Limestone	$CaCO_3$	1.60	2.71	4.43	10.5
Imported chalk	$CaCO_3$	1.60	2.71	4.43	12.5

Table 14-7 (cont.)

Pigment	Approximate Chemical Composition	Refractive Index	Specific Gravity	Bulking Value (gal/100 lb)	Oil Absorption (lbs oil/ 100 lb of pigment)
Calcium sulfate pigments					
Gypsum	$CaSO_4 \cdot 2H_2O$	1.53	2.35	5.107	21.0
Anhydrite	$CaSO_4$	1.59	2.95	4.070	25.0
Precipitated calcium sulfate	$CaSO_4$	1.59	2.95	4.070	50.0
Silicate pigments					
Silica	SiO_2	1.55	2.6	4.60	25.0
Diatomaceous silica	SiO_2	1.40-1.50	1.95-2.35	6.15-5.10	25.0-150.0
Clay	$Al_2O_3 \cdot 2SiO_2 \cdot 2H_2O$	1.56	2.60	4.60	36.0
Pyrophyllite	$Al_2O_3 \cdot 4SiO_2 \cdot 2H_2O$	1.588	2.85	4.22	
Talc	$2MgO \cdot 4SiO_2 \cdot H_2O$	1.59	2.85	4.212	27.0
Mica					
Phlogopite	$K_2O \cdot 6MgO \cdot Al_2O_3 \cdot 6SiO_2 2H_2O$	1.606	2.5-3.0	4.365	24.0
Muscovite	$K_2O \cdot 3Al_2O_3 \cdot 6SiO_2 \cdot 2H_2O$	1.59	2.5-3.0	4.272	47.5

Table 14-8

Tinting Strength and Hiding Power
of White Pigments

Pigment	Tinting Strength	Hiding Power (sq ft/lb)
Rutile titanium dioxide (PSC)	1850	157
Rutile titanium dioxide (conventional)	1750	147
Anatase titanium dioxide	1250	115
50% rutile calcium-base	880	82
Zinc sulfide	640	58
30% rutile calcium-base	600	57
Lithopone	280	27
Antimony oxide	300	22
Dibasic lead phosphite	250	20
Zinc oxide	210	20
35% leaded zinc oxide	175	20
Basic carbonate white lead	160	18
Basic sulfate white lead	120	14
Basic silicate white lead	80	12

Table 14-9

Colored Pigments

Medium chrome yellows are considered good.
Shading yellows - dirtier shade lemon yellows, are considered fair.

Zinc Yellows
   Zinc yellows are considered fair to good.

Basic Zinc Chromate
   Basic zinc chromate is not recommended for tints; recommended for primer use onl

Strontium Yellow
   Strontium yellow has very good lightfastness in tints.

Nickel Titanate Yellow
   Nickel titante yellow or titanium yellow has very good lightfastness in tints.

Nickel Azo Yellow
   This pigment has excellent lightfastness in tints.

Cadmium Yellow
   Excellent lightfastness in tints.

Yellow Iron Oxide
   Excellent lightfastness in tints.

Hansa Yellow
   Very good but recommended for interior use only.

Benzidine Yellow
   Fair and recommended for interior use only.

Vat Yellows
   Excellent lightfastness in tints.

Chrome Orange
   Light chrome orange is considered good but has a tendency to darken.
   Medium chrome orange is considered good but has a tendency to darken.
   Deep chrome orange is considered good but has a tendency to darken.

Molybdate Orange
   Molybdate oranges rated from fair to good.

Cadmium Orange
   Light cadmium orange considered excellent.
   Deep cadmium orange considered excellent.

"Mercadium" Orange
   Considered excellent for tints.

Benzidine Orange
   Rated fair and recommended for interior use only.

Dinitraniline Orange
   Rated good but recommended for interior use only.

Table 14-9 (cont.)

Vat Dye Orange
    Anthraquinone types have excellent fade resistance.

Chrome Greens
    Light chrome green considered good to fair.
    Medium chrome green considered good.
    Deep chrome green considered good.

Chromium Oxide
    Considered excellent for tints.

Hydrated Chromium Oxide
    Considered excellent for tints.

Copper Phthalocyanine Green
    Considered excellent for tints.

Organic Green Toners
    PTA and PMA have only fair durability and are recommended for interior
    use only.

Iron Blues
    Considered good to fair depending on treatment.

Copper Phthalocyanine Blue
    Considered excellent in any tint shade range.

Ultramarine Blue
    Considered poor to good depending on vehicle system.

Organic Blue Toners
    PTA and PMA blue toners recommended for interior use only; rated poor to
    fair.

Indanthrene Blue
    Considered excellent in tints of all ranges.

Carbazole Dioxazine Violet
    Considered excellent except in high bake white finishes.

Organic Violet Toners
    PTA and PMA violets not recommended for exterior use. Fair in tints for
    interiors.

Mineral Violet
    Not recommended for exterior. Excellent for interior finishes only.

Lithols
    Rated fair and for interior use only.

BON (β-Oxynapthoic)
    Rated as poor to fair in tints.

Toluidine Reds
    Rated as fair but would not recommend use for tints in white due to
    bleeding characteristics.

Para Reds
    Rated as poor for tints.

Table 14-9 (cont.)

Lithol Rubines
  Rated as good but for interior use only.

Chlorinated Para Red
  Rated as good but for interior use only.

Quinacridone Reds and Maroons
  Rated as excellent.

Red Iron Oxide
  Rated as excellent.

Cadmium Reds and Maroons
  Rated as excellent.

"Mercadium" Reds and Maroons
  Rated as excellent.

Red Lead
  Not recommended as a tinting color.

Thioindigo Reds and Maroons
  They can be rated from poor to very good depending on type. "Thiosafast" and "Thiofast" are rated good. Most others rated poor in light tints.

Arylide Maroons
  Rated as good for interior use only. Would not recommend as a tinting color of whites due to bleeding characteristics.

Siennas and Umbers
  Rated as excellent for tinting.

Table 14-10

Plasticizers Sanctioned by Food and Drug Administration[a]

Acetyl tributyl citrate

Acetyl triethyl citrate

p-tert-Butylphenyl salicylate

Butyl stearate

Butylphthalylbutyl glycolate

Dibutyl sebacate

Diethyl phthalate

Diisobutyl adipate

Diphenyl 2-ethylhexyl phosphate

Epoxidized soybean oil (max iodine value 6, min oxirane oxygen 6.0%)

Ethyl phthalyl ethyl glycolate

Glycerol monooleate

Monoisopropyl citrate

Mono-,di- and tristearyl citrate

Triacetin

Triethyl citrate

3-(2-Xenoyl)-1,2,-epoxypropane

---

[a]Federal Register (Feb. 18, 1966). 31 F. R. 2897 Title 21, Ch.1, Subchapter B, Part 121, Food additives Subpart E, p.1.

Flattening Agents. A flatting agent is an additive which reduces the gloss or angular sheen of the dried film.

Preservatives and Fungicides. Many micro-organisms can metabolize the organic materials in coatings. Protection against this is done with mercurial compounds, sulfur compounds, phenols, etc.

Wetting Agents. Coatings which must penetrate into capillaries and wet rough surface thoroughly are often formulated with wetting agents that promote good wettability of the liquid on the substrate.

Miscellaneous Additives. Many different types of materials are added to give specific properties. Examples include corrosion inhibiting agents, aluminum flake that changes the optical properties of the paint, metallic zinc which has corrosion protection properties, odor agents, fluorescent agents, conductivity aids, etc.

Section 15

Surface Preparation Prior to Application of the Coating

Surface Preparation

Many substrates are easily and well prepared with water, detergent or solvent washing. Although such techniques may be employed for metal preparation, the resultant surfaces are rarely adequate for more than the mildest interior environment. High pressure water blasting techniques, while adequately removing dirt, grease, old paint films, etc., will not dislodge tightly adherent oxide or millscale, unless sand is injected into the water stream. Water blasting may be better than grit or sand blasting in the removal of more resilient deposits.

Hand and power tool cleaning techniques (wire brushing, needle descaling, etc.) will eventually remove all tightly bonded contamination, but in practice, such techniques are far from thorough, too costly, and generally ineffective. Another serious drawback is that power tools can drive contamination further into the steel rather than remove it. Flame cleaning (where an oxyacetylene flame is played across the metal) is only effective for removal of loosely bonded materials (paint, loose scales and rust). The combustible materials are removed from the steel by burning, and the differential rates of thermal expansion between the noncombustible contamination and the steel causes its delamination. However, tightly adherent scale is not well removed by this technique and this method should be reserved for less demanding applications. Flame-cleaned steel should be cleaned of all debris after treatment and painted while still warm.

Theory

There is a two-fold objective (Fig. 15-1) in the surface preparation of reactive engineering metals like steel. First, the preparatory technique should remove not only loose material from the surface but also chemically bonded scales and oxide films that take up reactive sites on the metal. The removal of such materials exposes these reactive sites for subsequent reaction with either primer or pretreatment.

Second, good preparation should increase the surface area by "roughing up" or imparting some definite profile to the surface. In this way, the actual exposed surface area per unit of apparent area is dramatically increased and reactive sites per unit area are multiplied considerably. Thus, the opportunity for either primary or secondary valency bonding with the coating system is increased and adhesion is much improved.

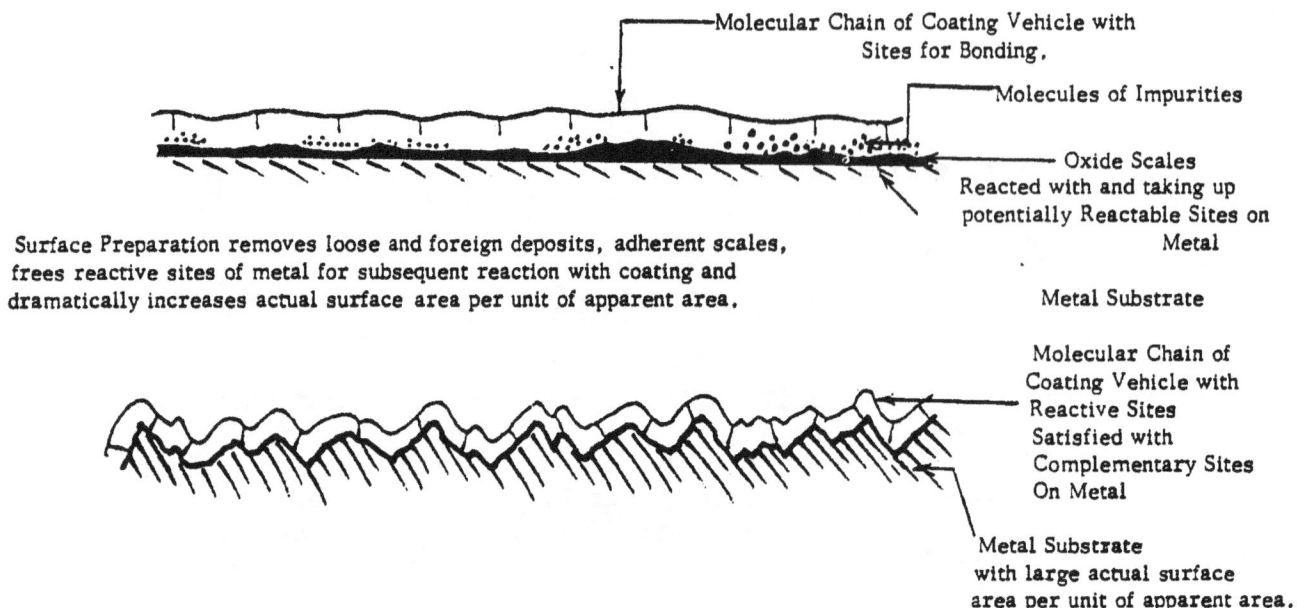

Molecular Chain of Coating Vehicle with Sites for Bonding.

Molecules of Impurities

Oxide Scales Reacted with and taking up potentially Reactable Sites on Metal

Surface Preparation removes loose and foreign deposits, adherent scales, frees reactive sites of metal for subsequent reaction with coating and dramatically increases actual surface area per unit of apparent area.

Metal Substrate

Molecular Chain of Coating Vehicle with Reactive Sites Satisfied with Complementary Sites On Metal

Metal Substrate with large actual surface area per unit of apparent area.

Figure 15-1. Surface Preparation

The best preparatory techniques are those that not only remove all contamination from the metal but also remove just enough metal to provide a scarified surface having an anchor pattern or "mil profile" of nearly microscopic peaks and valleys. These techniques should not remove too much metal which would be costly and may affect the mechanical properties of certain metallic parts. Such preparatory requirements are best fulfilled by either sandblasting or pickling. Sandblasting is preferred in the maintenance industry for preparation of steel for corrosive applications. Pickling, unlike blasting, is somewhat limited by the sizes of pickling tanks and is generally applied to small pieces as may be handled in industrial production lines.

Blast Cleaning

Blast-cleaned surfaces are classified according to the efficiency of the blast (Table 15-1). Surfaces having all contamination, scale, and rust removed and exhibiting the gray and white luster of virgin steel are classified as being "white blasted." These surfaces are intended for the most demanding requirements as may be expected in the chemical process industries or underwater environments. "Near white blasted surfaces" may exhibit traces (less than 5%) of the contaminate residues. Such surfaces are suitable for all environments except those involving actual immersion. The third category, the "commercial blast," is that most generally used in maintenance painting. Commercial blasted steel is free of all contaminants and two-thirds of its unit surface area is free of all visible residues. Slight discolorations and shadows (remaining on the steel after the removal of adherent rust and mill scale) are allowed over the other one-third of the surface. The least effective cateogry is the "brush off blast." This method removes all loose contamination, but not necessarily tight mill scale, rust, and adherent paint

Table 15-1

Techniques for the Preparation of Steel Surfaces Prior to Painting

Name	SSPC Designation	NACE Designation	Description
Solvent cleaning	SSPC-SP 1-63	None	Complete removal of oil, grease, wax, dirt and other contaminants by cleaning with solvents, vapours, emulsions alkalis or steam. Interior environments of low humidity only.
Hand tool cleaning	SSPC-SP 2-63	None	Removal of loose rust, loose mill scale and paint by manual labour with wire brushes, hand scrapers, sanders etc.
Power tool cleaning	SSPC-SP 3-63	None	Much the same as above but utilizing powered tools such as chippers, descalers, grinders etc.
Flame cleaning	SSPC-SP 4-63	None	High temperature flame dehydrates and removes rust, loose millscale and some tight millscale. Usually followed by wire brushing or blasting.
White metal blasting	SSPC-SP 5-63	NACE #1	Complete removal of all visible rust, paint, millscale, and foreign material by wheel or pressure blasting using (wet or dry) sand, grit or shot. Suitable for all the most severe environments including immersion service.
Commercial blast	SSPC-SP 6-63	NACE #3	Sandblasted until at least two-thirds of each element of surface area is free of all visible residues.
Brush-Off Blast	SSPC-SP 7-63	NACE #4	Blast cleaning of all except tightly adhering residues of millscale, rust and old coatings, exposing numerous, evenly distributed areas of underlying metal.
Pickling	SSPC-SP 8-63	None	Complete removal of rust and millscale by acid pickling, duplex pickling, or electrolytic pickling.
Weathering, followed by blast cleaning	SSPC-SP 9-63	None	All or part of the millscale is removed by allowing the steel to weather, followed by one of the blast cleaning standards.
Near white blast	SSPC-SP 10-63T	NACE #2	Blast clean until at least 95% of each element of surface area is free of all visible residues.

films. However, the entire surface must be subjected to the blast and numerous flecks of underlying metal must be visible uniformly across the surface. Brush off blasted surfaces are suitable for paint systems in mild environments where there is no exposure to immersion, chemicals, or abrasions.

The better the blast, of course, the more thorough the job must be, the longer it will take, and the higher will be the cost. Sand is not the only abrasive used for blasting. Crushed iron or steel grit, or iron or steel shot are also used. The abrasive size governs the depth of profile produced on the blasted surface (e.g. an 80 mesh sand gives a profile of about 1.5 mils and a 12 mesh sand produces a 2.8 mil profile). Some variation will be created by the type of steel, the impingement angle, and particle velocities. In most cases steel abrasive is recycled to reduce cost. This may also reduce the profile height as well as impede the efficiency of the blast. The profile height is the distance from the bottom of the lowest pits to the top of the highest peaks.

The required mil profile is dictated by the dry film thickness of the total paint system. The anchor pattern must be deep enough to hold the film securely, but not so deep that the peaks of the pattern are left with inadequate coating thickness. Thus a 1.25 mil profile is adequate for films where the total coating system does not exceed 8 mils, a 2 mil profile is recommended for thicknesses between 11 and 15 mils, and a 3.5 mil profile is used for coating films that exceed 15 mils. Mil profiles much below 1 mil are rarely used for anti-corrosive systems.

In practice, two techniques are employed in blast cleaning. In the first, the abrasive is injected into a pressurized air stream and expelled through a Venturi nozzle at 90-100 psi to impinge upon the surface being cleaned (Figure 15-2). This technique is much like conventional spray application. The second technique, centrifugal blasting, is done in large machines where the abrasive is fed into rapidly rotating wheels (Figure 15-3). These impart considerable velocity to the abrasive and fling it at the steel which is passed slowly through the machine.

Because of the health hazards of finely divided dust, the use of masks and special respirators is essential during the blasting process. Despite these precautions, difficulties may also arise from the contamination of surrounding air, leading to prohibition of sand blasting by some local authorities. The removal of old lead-based paint by blasting, for example, can create particulate lead levels in excess of 50 mg/m^3 in the immediate environment. Wet blasting, where water with suitable corrosion inhibitors (polyphosphates, chromates, etc.) is mixed with the abrasive, is often used here. Another mechanical approach to the same problem is the use of powerful suction devices that surround the nozzle and fit snugly against the object being blasted. In this way the abrasive may be removed immediately after impingement. Vacuum blasting is, however, very costly and time consuming.

After blasting care must be taken to ensure that no resulting contamination remains on the metal surface. All sand and metal dust should be vacuumed from the blasted surface.

Any newly prepared metallic surface is extremely reactive and must be coated as soon as possible after surface preparation. Steel will begin to corrode in milliseconds after blasting, although the initial corrosion products are thin, invisible, highly adherent films. In highly humid environments, the build-up of heavier loose corrosion products occurs very rapidly. All steel must be primed (or certainly pretreated) before these visible rust products appear. Even in moderate climates priming should be done within eight hours of exposure of virgin steel to the air. Radical surface preparation such as blasting should not be performed when ambient humidity is much above 80% or when there is significant risk of precipitation. Should a freshly blasted surface develop rust bloom before priming can take place, there is no alternative but to reblast the steel.

Figure 15-2. Air Pressure Sandblasting

(1) Compressor giving an adequate and efficient supply of air.
(2) Air hose, couplings and valves of ample size.
(3) A portable, high production sandblast machine.
(4) The correct size anti-static sandblast hose (with externally fitted quick couplings).
(5) High production Venturi nozzle.
(6) Pneumatic remote control valves for safety and cost savings.
(7) Moisture separator.
(8) High air.pressure at nozzle.
(9) Correct type and size of abrasive.
(10) Air fed helmet and air purifier (in good working order).
(11) A well trained operator.

Courtesy Clemco-Clementina Ltd.

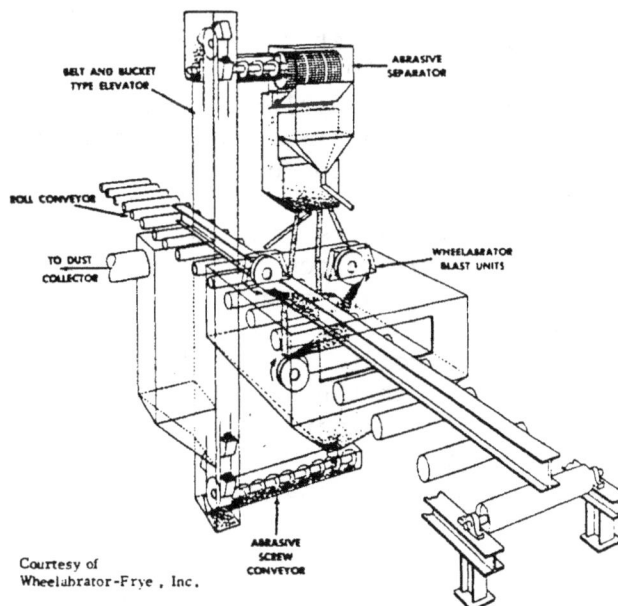

Figure 15-3. Schematic illustration of a typical centrifugal sandblasting unit.

Pickling

Pickling, the alternative method of successfully preparing a metal surface for painting, is normally done by immersing the work in the pickling fluid. Spray applications have been known, but are considerably less common than the immersion process. In pickling, the fluid reacts with the metal, the oxide scales, and the rust deposits, etching the steel and converting the oxides to soluble salts removed (together with excess fluid) by a subsequent rinsing in hot water.

Fluids may be proprietary in nature, but are most often aqueous solutions of acids. Caustic soda solutions and proprietary alkaline etchings are used for aluminum. Zinc and cadmium are etched in cold dilute sulphuric acid, although nitric acid may also be employed. Silver, when necessary, is etched with dilute nitric acid. Lead is pickled at room temperature in hydrochloric acid. (Vapor degreasing with chlorinated solvents is generally all the surface preparation lead will need). For copper and brass, a combination of sulphuric acid with either a nitric or chromic acid is used. Chromic acid tends to passivate copper surfaces forming impervious oxide films which reduce paint film adhesion. These can be removed after the chromic acid bath by a post-dip in nitric or sulphuric acid. Magnesium may be etched with either chromic or nitric acid. Dow Chemical Co., as a producer of magnesium, has developed a series of proprietary pretreatments to improve the paintability of this metal some of which (#1 and #20) are, in part, pickling treatments. The #1 (chrome pickle) utilizes sodium dichromate and concentrated nitric acid and is used for wrought magnesium. The #20, a modified chrome pickle used for magnesium castings, adds small amounts of an acid fluoride and a magnesium or aluminum sulphate to the bath.

Pickling of ferrous metals is accomplished by using a variety of agents. After removal of oils, etc., from the surface (usually by use of a caustic bath and rinse), sulphuric acid, diluted to 5-25% by weight from 140° to 180°F, hydrochloric acid at room temperatures at 25-50% by volume, and phosphoric acid at 15-40% by weight at 120° to 150°F, may be used effectively. In addition, other acids, such as nitric, hydrochloric, and even certain organic acids, have been used. In pickling steel, care must be taken to avoid hydrogen embrittlement. Not only does the absorption of atomic hydrogen embrittle the metal, but its release in molecular form (as a gas) after the metal has been coated (particularly during baking cycles) may result in blistering of the cured coating. Such out-gassing may be controlled by baking the steel before coating to a slightly higher temperature than that employed in the baking cycle. For example, martensitic stainless steels should be baked for at least 24 hours at 400°F after pickling in strong acids. Often the problem is controlled by the use of certain pickling inhibitors in the acid bath. Such inhibitors (used at about 0.1%) effectively control the extent of acid attack on the metal by adsorbing onto the metal surface as a mono-molecular layer. In this way, both the discharge of hydrogen ions and the dissolution of excess metal is controlled.

After pickling, it is important that all acids and salt residues from the pickling processes are removed from the metal before painting. This is normally performed by a water rinse immediately after the parts are removed from the pickling bath. Hot water rinses conveniently dry off the metal before rusting can occur. Metal rinsed in cold water baths are dried by compressed air, ovens, displacement of water with solvents, centrifuging, and a variety of other techniques.

Conversion Coatings and Metal Pretreatments

Phosphate Coatings

Steel pickled in phosphoric acid is not only etched but given a very thin phosphate coating. This in itself slightly improves the corrosion resistance of the metal and forms a porous primary bonded base for good paint adhesion. However, because the concentrations of the pickling solutions are high (15-40% by wt acid) a substantial film cannot be deposited. Less concentrated acid solutions containing zinc, manganese or alkali metal phosphates support the growth of thicker crystalline films of tertiary (and sometimes secondary) phosphate salts. These give better corrosion resistance (including good resistance to underfilm creepage) and a very sound substrate for the subsequent adhesion of organic coating films.

Typically, a phosphatizing reaction proceeds as is described in the following illustration using a primary zinc phosphate (mono-zinc orthophosphate)/ phosphoric acid system for the treatment of steel. At the elevated temperatures at which the bath is maintained, the primary (acid) phosphate is converted to a tertiary phosphate. At this point, the bath concentration must be controlled so that the tertiary zinc phosphate is below the saturation point of the bath. (If manganese phosphate is used, some secondary manganese phosphate will also form).

$$3Zn(H_2PO_4)_2 \longrightarrow Zn_3(PO_4)_2 + 4H_3PO_4$$

primary zinc phosphate   tertiary zinc phosphate

When the steel is dipped into the coating fluid, it is attacked by phosphoric acid which liberates hydrogen at the cathode and forms primary and secondary iron phosphates.

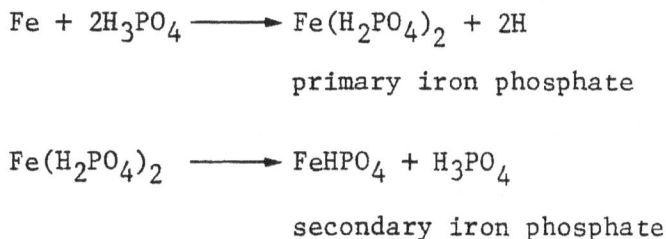

$$Fe + 2H_3PO_4 \longrightarrow Fe(H_2PO_4)_2 + 2H$$

primary iron phosphate

$$Fe(H_2PO_4)_2 \longrightarrow FeHPO_4 + H_3PO_4$$

secondary iron phosphate

At the immediate metal surface this reaction increases the pH of the solution. Consequently, the less soluble phosphates (tertiary zinc and, to a lesser extent, secondary iron) are precipitated as a fine crystalline layer on the steel. The overall reaction may be written:

$$Fe + 3Zn(H_2PO_4)_2 \longrightarrow Zn_3(PO_4)_2 + FeHPO_4 + 3H_3PO_4 + H_2$$

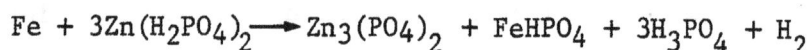

During this treatment the active ingredient of the bath will undoubtedly become depleted. Too much sludge is formed if the acid content falls too low; on the other hand, if the bath becomes too acidic, metal loss from the steel will be excessive and phosphate films will be thinner. Care must be taken, therefore, to adjust the bath concentration in close conformance with the phosphate coating manufacturer's instructions.

In practice, the basic reaction is so greatly retarded by hydrogen polarizing the cathode that efficient de-polarizers are added in the form of nitrates, nitrites, and other oxidizing agents. In this way an unmodified reaction that might take hours to develop an adequately thick film is accelerated to form the same film in a minute or so. The actual compositions of the phosphate processing solutions are often proprietary and their reactions quite complex. The patent literature on these materials is extremely extensive. In addition to the zinc, manganese, and alkali metal phosphates indicated above, ammonium acid phosphates, arsenic, molybdenum and copper salts, and tannins have all been employed with interesting results. In addition to their use on steel surfaces, zinc phosphate coatings are widely used as conversion coatings for zinc, aluminum, and cadmium. Other specific compositions are available for these other metals and alloys.

Phosphatizing is a long-standing and widely accepted technique, used for many industrial line applications including automotive parts and household appliances. In practice, phosphatizing involves a multi-stage process. First the steel is degreased in chlorinated solvents or a hot alkaline bath. After a rinse stage, the steel is phosphatized, rinsed again, and finally treated with a very dilute solution (0.05%) of chromic acid. The rinse probably converts the tetrahydrated tertiary zinc phosphate (deposited from the bath) to the dihydrate which improves the overall corrosion resistance of these systems. Application is usually made by immersion into phosphate baths at 160-180°F, although spray applications are also possible. Spray applications are usually made at lower temperatures (140°F). Some proprietary treatments can be carried out at temperatures not much above room temperature.

Generally, zinc phosphate coatings are thicker (0.05- 0.4 mils) than coatings of "iron phosphate" (0.005-0.03 mils). Iron phosphate coatings are not generally considered as good as zinc phosphate coatings but they are cheaper, and their thinner films give, perhaps, a better appearance. Control of crystal size in the deposition of both coatings is most important, protection decreasing as crystal size increases. Crystal size can be regulated with temperature and/or suitable additives. Immersion in propriety solutions containing titanate salts is also valuable in controlling the crystal size. Phosphate coatings are quite porous and begin to degenerate shortly after application if not quickly recoated with a suitable primer. In less than four days the unprotected iron phosphate film will deteriorate to a point where there is little corrosion protection.

Phosphate coatings may be recoated with a variety of organic finishes including alkyds, epoxies, acrylics, and amino baking finishes. They also form a good substrate for subsequent electrophoretically applied coatings.

Chromate Conversion Coatings

    Chromate conversion coatings, while not employed on steel, are used in
treating zinc, aluminum, tin, and magnesium. These coatings are not as valuable for increasing paint adhesion as the phosphate conversion coatings, but
often give a better surface for subsequent paint adhesion than virgin metal.
Chromate coatings also convey some corrosion resistance to these surfaces.

    As with phosphate coatings, the number of proprietary chromate coatings
is very extensive. In the British M. B. V. process for aluminum, the metal is
immersed for a few minutes in an alkaline chromate bath (a dilute solution of
sodium chromate and sodium carbonate). The similar Alrok Process uses potassium
dichromate in lieu of sodium chromate. The Alodine Process employs a bath containing chromic, phosphoric, and hydrofluoric acids and deposits gray-green or
golden brown, slightly iridescent coatings of aluminum phosphate, chromium
phosphate, and aluminum oxide.

    Newer chromate treatments for aluminum contain both soluble hexavalent
and insoluble trivalent chromium materials with fluorides (occasionally chlorides and sulphates) as activators. These give thinner films than the Alodine
Process (0.02 mils as opposed to 0.1 - 0.4 mils). Film color varies with increasing thickness of the deposits, thin films being iridescent and almost
colorless, thicker films becoming yellow to golden and finally brown. Film
thickness is directly related to immersion time, 20 seconds being necessary
for the thinner deposits, while less than five minutes produce the browner
films. Baths are not difficult to maintain and are controlled by pH adjust-
ments often artificially made.   pH is normally kept between 1.2 and 2.2, although ferricyanide additons have been used to expand this range. Below this
range, however, the deposited conversion coatings become too powdery, above it
they become too thin. In some processes, nitric acid and caustic soda additions are used for pH control. Control of the chromate and fluoride is done
separately.

    Similar chromate treatments are used on zinc and cadmium. Again, both
hexavalent and trivalent chromium compounds are used in the bath. Activators
may be more varied. In addition to fluorides, chlorides, and sulphates, their
parent acids and such organic acids as the lower carboxylic acids may be used.
Coatings range in color from slightly iridescent (0.02 mils) to bronze (0.3
mils), the latter coatings being thicker and more protective. Baths are maintained at a pH of 1.0 to 3.5. Like the chromate coatings on aluminum, these
coatings are initially soft but after drying become hard and impermeable to
water.

    While aluminum surfaces are preferably etched in alkaline materials
before treatment with chromate, galvanized metal may be alternatively treated
in phosphoric acid. Zinc and cadmium die castings are normally dipped in
dilute sulphuric acid. Chromate conversion coatings are also applicable for
use on silver surfaces.

    For brass (except leaded brass) and copper surfaces, similar treatments
yield iridescent yellow conversion coatings. These are useful as paint bond
treatments where pigmented organic films are specified. Copper and brass are
usually coated with clear acrylics, etc. (often inhibited with benzotriazole).
For these systems, the metal is given a "bright dip" using a combination of

sulphuric and nitric or chromic acids.

Some of the many pretreatments for magnesium developed by the Dow Chemical Co. might be classified as chromate treatments. Applied after an alkaline etch and a subsequent rinse, treatment #1, often referred to as a chrome pickle, uses a sodium dichromate/nitric acid bath followed (after rinsing) by a dichromate bath. In all, there are six principal preparatory treatments for magnesium (#1, #7, #9, #10, #17, and #19) involving four to seven steps each. Treatment #17 is an electrochemical type (anodizing) giving abrasion resistant films of excellent corrosion resistance. For more detailed information the book "Magnesium Finishing" by the Dow Chemical Co. is highly recommended.

Detailed discussion of the surface treatment of the miscellaneous metals is beyond the scope of this unit, but Table 15-1 may be used as a rough guide to the surface preparation necessary for paint adhesion.

Table 15-1

Guide to the Surface Preparation and Pretreatment
of the Miscellaneous Metals

Quality of Adhesion vs. Type of Surface Treatment

Metal	Solvent Cleaning	Abrasive Cleaning	Etching	Chemical Conversion Coating
Beryllium	Poor	Good	Good	--
Beryllium-copper	Poor	Good	--	--
Cadmium	Poor	Fair	--	Excellent
Chromium	Poor	Good	Good	--
Copper, brass & bronze	Poor	Poor	Poor	Excellent
Galvanized metal	Poor	Good	--	Excellent
Gold	Good	Good	--	--
Iron, cast	Poor	Good	Good	Excellent
Lead	Excellent	Excellent	--	--
Nickel plate	Poor	Good	Good	--
Nickel-base alloys	Poor	Good	Good	--
Pewter	Poor	Good	--	--
Platinum	Good	Good	--	--
Rhodium	Good	Good	--	--
Silver	Good	Good	--	Excellent
Solder (tin/lead)	Excellent	Excellent	--	--

Table 15-1 Cont'd.

Quality of Adhesion vs. Type of Surface Treatment

Metal	Solvent Cleaning	Abrasive Cleaning	Etching	Chemical Conversion Coating
Terne plate	Poor	Good	--	--
Tin	Poor	Good	--	--
Tungsten & tungsten carbide	Poor	Fair	Good	--
Uranium (depleted)	Poor	Good	--	--
Zinc	Fair	Good	Good	Excellent

Other Treatments

Various other pretreatments have been examined, and some have been used in various parts of the world. These include oxide and oxalate coatings, principally on steel and stainless steel. These are used in the United States by the military on weapons, etc. Oxide (magnetite) film forming processes for steel, utilizing strong caustic solutions with suitable oxidizing agents are reported to give good surfaces for painting. Similar black coatings, having minimal light reflectance properties serve functional purposes in optical applications, and could lend themselves to use on solar energy collectors, etc.

Oxide coatings on aluminum are widely used. Although the natural oxide layer on this metal is not a good substrate for paint, it is possible, after degreasing and cleaning the metal, to chemically remove this coating and artificially re-form an oxide layer under closely controlled conditions. This treatment is known as anodizing. In the anodizing process, the clean aluminum to be treated is made the anode of an electrochemical cell employing an electrolyte of either sulphuric, chromic, phosphoric, oxalic, or boric acid, depending upon the type of coating required. Boric acid is rarely used for prepaint treatments as the oxide deposited in this case is thin, hard, and nonporous. The other acids give thicker, more porous films. The effect of the electrolyte is apparently one of solvency; where boric acid has no solvency for the oxide a thin continuous coating is formed. The stronger acids, having some solvency for the oxide, continually remove a percentage of the oxide as it forms, leaving a porous but thicker film.

In practice, 15% sulphuric acid is the most widely used electrolyte. The bath is operated at about 95°F, with a current density of approximately 12 amps per ft^2 and a voltage of 10-20 volts. Increasing the acid concentration, decreasing the voltage and current density, and raising the temperature all tend to produce more porous coatings.

Chromic acid baths, also common, employ lower acid concentrations (about 10%) with lower current densities (2.5 to 4 amps per ft^2) and higher voltages (about 40 volts). Temperatures are much the same as above.

Certain proprietary alkaline oxidizing compounds for blackening copper and brass have also been found to improve the subsequent adhesion of styrenated alkyd paints.

Wash Primers

Although field application of phosphate coatings on sandblasted steel prior to the use of inhibitive metals primers is not unknown, the principal pretreatment or metal conditioner of the maintenance painting industry is the WP-1 Wash Primer. This unique material is neither a conversion coating nor an organic metal primer in the strict sense, rather it displays characteristics of both. Certainly it forms primary valency bonds with the base substrate (steel, aluminum, zinc, tin, stainless steel, or titanium) and its reactions include the formation of both phosphates and chromates in situ. Equally true, however, is the presence of an organic binder (polyvinyl butyral), while the chromates and phosphates formed certainly act in many ways as inhibitive pigments.

A typical composition for the wash primer is found in Table 15-2. The acid component must be added slowly and carefully to the base under agitation immediately before use. In this way the reaction is initiated. The in-can reaction (even in the absence of metal) will continue for some eight hours after which the reactants will be exhausted and the material useless. The WP-1 Wash Primer does not gel (as do two-pack thermosets), but nonetheless, the pot life is definitely limited.

Table 15-2

WP-1 Wash Primer

Percentage by Weight	
Base Component	
Vinyl butyral resin	7.20
Zinc tetroxychromate.	6.93
Magnesium silicate.	1.11
n-Butanol	16.06
Isopropanol (99%)	48.70
	80.00
Acid Component	
Phosphoric acid (85%)	3.60
Isopropanol (99%)	13.20
Water	3.20
	20.00

Single component systems (Table 15-3) have been developed which utilize less soluble inhibitors such as lead chromate and chromic phosphate. Although these materials have gained some acceptance, they are not nearly so widely used as the two-component systems. This is perhaps because of their rather more specific utility compared to the wide spectrum of applications that the two-package varient may satisfy. Their reaction mechanism may also be different.

Table 15-3

Single Package Wash Primers

	11A	11B	11C	11D	11E
	Percentages by Weight				
Vinyl butyral resin	9.0	9.0	9.0	9.0	9.0
Lead chromate (low solubility)	8.6	--	--	--	--
Chromium phosphate (salt free)	--	9.0	4.5	9.0	4.5
Zinc tetroxychromate	--	--	4.5	--	4.5
Magnesium silicate	1.4	1.4	1.4	1.4	1.4
Ethanol or isopropanol (99%)	53.0	54.5	54.5	54.5	54.5
Methyl isobutyl ketone	13.0	16.1	16.1	16.1	16.1
Phosphoric acid (85%)	2.9	0.9	0.9	1.8	1.8
Ethanol	9.2	8.2	8.2	6.4	6.4
Water	2.9	0.9	0.9	1.8	1.8
	100.0	100.0	100.0	100.0	100.0

The WP-1 Wash Primer should be applied to clean metal in a very thin film (0.3 - 0.5 mils); thicker films may delaminate. In the presence of metal the reactions occurring in the film are thought to proceed as follows: the phosphoric acid converts the zinc tetroxychromate to chromic acid, zinc phosphate, and other chromates of lower basicity. The primary alcohol is then oxidized by the chromic acid to its respective aldehyde and, in the presence of more phosphoric acid, chromic phosphate is formed. Still, more phosphoric acid attacks the metallic substrate, depositing a film of the pertinent phosphate. At the same time, the chromium phosphate forms a chelated matrix with the polyvinyl butyral resin creating a complex film that is bonded to the deposited phosphate coating by primary and/or secondary valency bonds. Free chromates and phosphates remaining in the film serve as a source of inhibitive ions in the same manner as do the inhibitive pigments of an organic metal primer.

Although the wash primer has been in wide use for over 30 years, mistakes are still made in its application. The mixing ratio (4 parts base to one part acid by volume) is relatively critical as is the speed of incorporation. Addition of one component too rapidly to the other will produce gelation. Acid pickled surfaces are not recommended for wash primer application because of the tendency of pickling agents to react preferentially with the metal, satisfying groups on the metal that might have reacted with the wash primer. Moreover, acid salts not removed after pickling cause serious osmotic blistering in wash primed systems for eventual fresh water immersion. Paradoxically, the wash primer has good adhesion to surfaces that have been phosphatized and even those treated with a chromic/phosphoric rinse.

The wash primer may be applied by most methods. It should be applied as a wet film to ensure complete wetting of the substrate so that the subsequent reactions can take place. Spray application may be facilitated by diluting the mixed material with more alcohol. On damp surfaces, application may be successful by thinning with a little n-butanol to prevent resin precipitation by the moisture from the substrate.

The WP-1 Wash Primer serves as a pretreatment for many coating systems in many environments. Alkyds (including styrenated alkyds), epoxies, phenolics, urethanes, amino-baking systems, emulsion paints, and even many chlorinated rubber-based systems adhere to the wash primer. Acrylics and nitrocellulose lacquers, in general, do not. Vinyl chloride/vinyl acetate lacquers, although widely used as top coats, must be vinyl alcohol modified. Unmodified vinyl lacquers show poor adhesion to wash primed surfaces. Vinyls modified with maleic acid give intermediate results.

Vinyl systems over wash primed steel have been used widely in marine applications on ships, off-shore oil facilities, piers, and other structures. In fresh water, particularly at temperatures above 100°F, such systems are not as reliably effective because of a tendency toward osmotic blistering, although use has been made of the wash primer in all vinyl systems for fresh water immersion. Vinyl finish coats pigmented with leafing aluminum do much to reduce osmotic blistering of wash primer systems in fresh water. In general, however, such fresh water vinyl systems forego the use of the wash primer and rely on the maleic acid modified resin without inhibitive pigment as a simple barrier system. Because of the maleic acid reactivity, the aluminum is packaged separately and mixed into the resin just prior to painting.

Proper surface preparation is important in providing a foundation for the protective coating system. There are various methods and locations for performing the surface preparation function. Decisions in establishing the project specification must weigh standards of quality as well as the economics of surface preparation, which can amount to 50% of total coating costs.

The surface preparation required for different types of coating systems to be applied over structural steel vary considerably, depending on the type of coating as well as the service environment. The Steel Structures Painting Council, the National Association of Corrosion Engineers, the American Water Works Association, and the American Society for Testing and Materials all make reference to various standards that define degrees and methods of surface preparation. The most widely used surface preparation specifications are those published in STEEL STRUCTURES PAINTING MANUAL, Vol. 2, Systems and Specifications. (See Appendix G.) Given below is a brief description of the various SSPC Surface Preparation Specifications as they apply to structural steel. It is recommended that the reader refer to the SSPC Manual for the complete text of the specifications:

SSPC-SP1 Solvent Cleaning:

The removal of dirt, oil, grease, and foreign matter is accomplished with solvents or commercial cleaners using various methods of cleaning such as wiping, dipping, steam cleaning, or vapor degreasing.

It is generally conceded that solvent wiping does not positively remove all oil and grease from the surface. Therefore, a more efficient cleaning method, such as vapor degreasing or steam cleaning, should be employed where coatings will not tolerate any oil or grease residue.

SSPC-SP2 Hand Tool Cleaning:

The removal of loose rust and mill scale is carried out by hand wire brushing, scraping, chipping or sanding. Hand tool cleaning does not remove all rust residue or intact, firmly adhering mill scale.

SSPC-SP3 Power Tool Cleaning:

The removal of loose rust and mill scale is accomplished by mechanical means such as power sanders, wire brushes, chipping hammers, abrasive grinding wheels, or needle guns. Power tool cleaning provides a slightly higher degree of cleanliness than does hand tool cleaning but is not regarded as adequate surface preparation for long-term exterior exposure of most high performance coating systems.

SSPC-SP5 White Metal Blast Cleaning (NACE 1):

The complete removal of all visible rust, mill scale, paint, and foreign matter is accomplished by compressed air nozzle blasting, centrifugal wheels, or other specified method, leaving an overall, uniformly gray-white metallic appearance.

SSPC-SP6 Commercial Blast Cleaning (NACE 3):

The removal of at least two-thirds of all visible rust, mill scale, paint, and other foreign matter from each square inch of surface is accomplished by compressed air nozzle blasting, centrifugal wheels, or other specified method.

SSPC-SP7 Brush-off Blast Cleaning (NACE 4):

The removal of loose rust, mill scale, paint, and foreign matter from the surface is carried out by compressed air nozzle blasting, centrifugal wheels, or other specified methods.

SSPC-SP8 Pickling:

The complete removal of all rust, mill scale, and foreign matter is accomplished by chemical reaction or electrolysis in acid solutions. The degree of cleanliness is similar to SSPC-SP5 White Metal Blast Cleaning.

SSPC-SP10 Near-White Metal Blast Cleaning (NACE 2):

The removal of 95% of all visible rust, mill scale, paint, and other foreign material from each square inch of surface is accomplished by compressed air nozzle blasting, centrifugal wheels, or other specified method.

Degree of Cleanliness vs. Coating Performance:

Abrasive blast cleaning, as defined in SSPC Specifications SP5, SP6 and SP10, is often regarded as the preferred method of surface preparation for carbon steel. Experience has proven that any coating system, applied over a properly blast-cleaned surface, will cost less per square foot per year than the same system applied over hand or power tool cleaned surfaces.

GUIDELINES FOR INSPECTION OF SURFACE PREPARATION

1.   Check surface contamination prior to cleaning, bearing in mind expected methods of cleaning.

2.   Check condition of welds for spatter, slag, and sharp protrusions.

3.   Check abrasive visually for contaminants.

4.   Check air supply to blast nozzle for contaminants such as oil and water.

5.   Check degree of blast cleanliness for conformance with standard.

6.   Check blast profile for conformance with standard.

7.   Check blasted surface for excessive surface dust and/or surface contaminants.

5

Section 16

Methods of Applying an Organic Coating

Organic coatings are applied in many ways. Coatings based on solvent loss are commonly applied by dipping, brushing, rolling or spraying. Solvent loss may be hastened with some coatings by a thermal treatment, whereas in other coatings air-drying is necessary for development of optimum properties. Coatings based on dry powders containing no solvent are formed by causing the particles to flow by heating the substrate metal. Such coatings are used on reinforcing rods and on underground pipelines. Coatings may also be applied electrochemically in which a dispersion of an organic material in a conducting liquid is caused to precipitate at either the anode or the cathode because of the pH at the electrode or chemical reactions that take place at the electrode. Electrocoated materials must be cured and compacted in a second step. These coatings have found wide use as undercoats in automotive applications. Coatings may also be applied as laminates in roll forming operations or as wound tape as used on underground pipelines.

The most common method for the application of coatings to structures and equipment *in situ* is by spray painting. The paint is dispersed as a fine spray which clings to the target and then spreads as a thin film by surface tension forces. The spray may be formed by air atomization or by airless spray in which the spray is formed by forcing the material under high pressure through a small orifice. Airless spraying has the advantage of more uniform distribution of the paint, especially at corners and crevices. It has the disadvantage that care must be taken to prevent the formation of too high a void volume. Coatings may also be applied electrostatically by grounding the structure to be painted and charging the particles so that they are attracted to the substrate.

Corrosion protective coatings may be classified in terms of the binder type. Four general types may be identified: solvent evaporation types, oxidation or drying oil types, polymerization or co-reacting types, and inorganic silicate binder types. Information outlining the properties of such types of coatings is given in Tables 16-1, 16-2, 16-3, and 16-4.

Seven different general types of protective coatings may be identified. Preparation method and coating system are described under each system.

(1) Maximum durability in severe industrial service environments requires a 2- or 3-coat system:

> abrasive blast (SSPC - SP-10)
> 2.5 mils inorganic zinc-rich primer
> 5 mils high build epoxy polyamide (or high build vinyl)
> 2 mils aliphatic urethane or acrylic topcoat.

Table 16-1

Solvent Evaporation Types

	Petroleum Hydrocarbon	Vinyl Coal Tar	Chlorinated Rubber	Chlorinated Rubber, Oil or Alkyd Modified	Vinyl Alkyd	Vinyl Chloride Copolymers	Polyvinyl Chloride	Vinyl-Acrylic	Acrylic
I. Effect of sunlight	Minor	Slow surface chalk	Slow surface chalk	Slow surface chalk	Slow surface chalk	Slow surface chalk	Slow surface chalk	Very slow surface chalk	Prolonged chalk resistance.
II. Wet or humid environments	Excellent	Excellent	Excellent	Fair to good; will yellow	Good; slight yellowing	Excellent	Excellent	Very good	Good
III. Industrial atmosphere contaminants									
a) Acid	Excellent	Excellent	Excellent	Fair to good	Good	Excellent	Outstanding	Very good	Good
b) Alkali	Excellent	Excellent	Excellent	Fair	Fair	Excellent	Excellent	Very good	Good
c) Oxidizing agents	Good	Limited	Excellent	Fair	Fair	Excellent	Outstanding	Good	Good
d) Solvents	Poor	Poor	Limited	Limited	Limited	Limited	Good	Limited	Limited
IV. Spillage & splash of industrial compounds									
a) Acids	Good	Good	Very good	Not recommended	Not recommended	Very good	Excellent	Fair	Poor to fair
b) Alkali	Good	Good	Very good	Not recommended	Not recommended	Very good	Excellent	Fair	Poor to fair
c) Oxidizing agents	Fair	Not recommended	Good	Not recommended	Not recommended	Good	Excellent	Fair	Poor to fair
d) Solvents	Not recommended	Not recommended	Not recommended	Not recommended	Not recommended	Not recommended	Limited	Not recommended	Not recommen
V. Physical properties									
-Abrasion resist.	Poor	Good	Good	Fair	Fair	Good	Excellent	Good	Good
-Heat stability	Softens	Limited	Limited	Fair	Fair	Limited	Limited	Limited	Limited
-Hardness	Soft	Good	Good	Fair-good	Fair	Good	Good	Good	Good
-Gloss	None	None	SG to matte	Wide range	Limited range	SG to matte	Gloss to matte	Semi-gloss	Gloss, Semi-gloss
-Colors	Black and Aluminum	Black/Gray	Wide range	Full range	Full range	Full range	Most colors	Full range	Full range
VI. Additional notes	Water resistance & ease of repair result in wide use as ship bottom systems	Use as marine primer	Traditionally low solids		Useful as primer or top-coat for more chemically resistant vinyl copolymers	Available in high and low build formulas. High build more popular in maintenance applications.	Low build. Economic limit to tank lining & other severe exposures.	Wide use as finish coat with vinyl systems.	Use as finish coat for epoxy systems for improved weathering.

Table 16-2

Oxidation or Oil Types

	Silicone Alkyd	Epoxy Ester	Alkyd	Uralkyd
I. Effect of sunlight	Slow surface chalk	Rapid surface chalk	Slow surface chalk	Yellowing & surface chalk
II. Wet or humid enviroments	Fair to good	Fair to good; will yellow	Poor to good; will yellow	Fair
III. Industrial atmosphere contaminants a) Acid b) Alkali c) Oxidizing d) Solvents	Fair to good Fair Fair Limited	Fair Fair Poor Limited	Fair to good Poor Fair Limited	Poor to fair Fair Poor Good
IV. Spillage & splash of industrial compounds a) Acids b) Alkalis c) Oxidizing d) Solvents	Not recommended Not recommended Not recommended Not recommended	Not recommended Not recommended Not recommended Not recommended	Not recommended Not recommended Not recommended Not recommended	Not recommended Not recommended Not recommended Not recommended
V. Physical properties -Abrasion resistance -Heat stability -Hardness -Gloss -Colors	Fair Good Fair Good Limited	Fair Fair Good Wide range Full range	Fair Fair Fair Wide range Full range	Excellent Good Excellent High gloss Limited range

Table 16-3

Polymerization or Co-reacting Types

	Coal Tar Epoxy	Epoxy, Amine Cure	Epoxy, Poly-amide Cure	Urethane Moisture Cure	Urethane, Two Package
I. Effect of sunlight	Surface chalking	Yellowing & surface chalking	Yellowing & surface chalking	See Additional notes	See Additional notes
II. Wet or humid environments	Excellent	Very good - may yellow	Very good - may yellow	Very good	Very good
III. Industrial atmosphere contaminants					
a) Acid	Excellent	Good	Good	Good	Very good
b) Alkali	Excellent	Excellent	Excellent	Good	Excellent
c) Oxidizing	Limited	Limited	Limited	Poor	Limited
d) Solvents	Limited	Excellent	Excellent	Excellent	Excellent
IV. Spillage & splash of industrial compounds					
a) Acids	Good	Fair	Poor-fair	Fair	Good
b) Alkalis	Good	Excellent	Excellent	Fair	Good
c) Oxidizing agents	Not recommended	Not recommended	Not recommended	Not recommended	Not recommended
d) Solvents	Not recommended	Excellent	Very good	Good	Excellent
V. Physical properties					
-Abrasion resistance	Limited	Good	Good	Excellent	Outstanding
-Heat stability	Excellent	Good	Good	Good	Good
-Hardness	Very hard	Very hard	Hard	Excellent	Excellent
-Gloss	None	Wide range	Wide range	Range	Range
-Colors	Black, red	Full range	Full range	Limited range	Full range
VI. Additional notes	Timing between applications critical in obtaining intercoat adhesion.	Cure rate & pot life affected by temperature (True of all epoxies)		Have ability of curing at low temperatures. Uretha based on aromatic diisocyanates discolor & chalk rapidly in sunlight. Tho based on aliphatic diisocyanates have prolonged resistance to yellowing & surface chalking. Care must be taken to avoid moisture contact before cure.	

Table 16-4

Inorganic Silicate Binders Types

	Inorganic Zinc Post-Cured	Inorganic Zinc Self-Cure, Water-Based	Inorganic Zinc Self-Cure, Solvent-Based	Organic Zinc One Package	Organic Zinc Two Package
I. Cure type	Inorganic	Inorganic	Hydrolyzable organic silicate	Lacquer	Coreacting
II. Effect of sunlight	Unaffected	Unaffected	Unaffected	Surface chalking	Surface chalking
III. Wet or humid environments	Outstanding	Outstanding	Outstanding	Very good	Very good
IV. Industrial atmosphere contaminants a) Acid	Requires topcoat	Requires topcoat	Requires topcoat	Requires topcoat	Requires topcoat
b) Alkali	Requires topcoat	Requires topcoat	Requires topcoat	Requires topcoat	Requires topcoat
c) Oxidizing	Requires topcoat	Requires topcoat	Requires topcoat	Requires topcoat	Requires topcoat
d) Solvents	Outstanding	Outstanding	Outstanding	Limited	Excellent
V. Spillage & splash of industrial compounds a) Acids	Not recommended	Not recommended	Not recommended	Not recommended	Not recommended
b) Alkalis	Not recommended	Not recommended	Not recommended	Not recommended	Not recommended
c) Oxidizing	Not recommended	Not recommended	Not recommended	Not recommended	Not recommended
d) Solvents	Outstanding	Outstanding	Outstanding	Limited	Very good
VI. Physical properties -Abrasion resistance	Outstanding	Outstanding	Outstanding	Good	Good
-Heat stability	Outstanding	Outstanding	Outstanding	Good	Good
-Hardness	Outstanding	Outstanding	Outstanding	Good	Good
-Gloss	None	None	None	Flat	Flat
-Colors	Gray or tints of gray	Gray or tints of gray	Gray or tints of gray	Gray or tints of gray	Gray or tints of gray
VII. Additional notes	As primers for organic systems, provide greatly extended service life. Special application technique or use of tie coat may be required to avoid solvent bubbling in organic topcoat. Alkyds always require tie coat.			Can be used to touch up inorganic primers compatible with topcoats.	Choice of topcoat is critical.

(2) An alternate system giving slightly less corrosion resistance, especially at coating holidays or damaged areas:

> abrasive blast (SSPC-SP-6 or SP-10)
> 2 mils inhibitive epoxy primer
> 5 mils high build epoxy polyamide
> 2 mils aliphatic urethane or acrylic topcoat.

(3) Where aesthetic considerations are not of great importance, the aliphatic urethane or acrylic topcoat may be deleted without adversely affecting performance of systems (1) and (2).

(4) For previously painted surfaces where abrasive blasting is not permitted, the following is satisfactory:

> power tool clean and solvent clean
> 3 mils universal synthetic resin "universal" prime
> 5 mils high build vinyl or epoxy

Alternatively, where a minimum number of steps is required for both surface preparation and coating, a single coat of epoxy mastic can be applied in lieu of primer and high-build topcoat. The mastic epoxy coatings are often formulated with aluminum for added protection and heat reflective properties and can be applied in one step as an 8 mil coating.

(5) Where service conditions are relatively mild or protection is required for a short time only, alkyd or latex systems may be used.

> power tool and solvent clean
> 1.5 mils of zinc chromate alkyd primer
> 2 mils alkyd topcoat
> 2 mils alkyd topcoat.

(6) Organic binder systems where fire protection is afforded for up to 2 hours by intumescing or ablative properties. In the case of steel subject to a fire the goal is to prevent the steel from exceeding 1000°F where it loses 50% of its mechanical strength.

(7) For the protection of concrete, epoxies and urethanes are preferred.

> abrasive blast or acid etch
> epoxy primer/sealer
> epoxy monolithic topping (30-100 mils)

The approximate lifetimes of the above systems are:

System	Severe Service	Normal Industrial
1	5 - 15 years	10 - 20 years
2	5 - 10	10 - 15
3	5 - 15	10 - 20
4	2 - 5	5 - 15
5	½ - 2	2 - 4

Section 17

Primers, Mid Coats and Topcoats

No notes for this section. Subject will be very briefly covered during the class sessions.

Section 18

Compatibility of Organic Coatings

Table 18-1

Generic Compatibility of Coatings

PRIMER OR EXISTING COATING	OIL DRYING			LATEX		SOLVENT DRYING		CHEMICALLY REACTING			
TOPCOAT	Oleoresinous	Alkyd	Silicone Alkyd	Acrylic	Polyvinyl Acetate	Vinyl	Chlorinated Rubber	Epoxy	Coal Tar Epoxy	Urethane	Polyester
OIL DRYING											
Oleoresinous	C	C	C	CT	CT	NR	NR	NR	NR	NR	NR
Alkyd	C	C	C	CT	CT	NR	NR	NR	NR	NR	NR
Silicone Alkyd	C	C	C	CT	CT	NR	NR	NR	NR	NR	NR
LATEX											
Acrylic	C	C	C	CT	CT	C	C	NR	NR	NR	NR
Polyvinyl Acetate	C	C	C	CT	CT	C	C	NR	NR	NR	NR
SOLVENT DRYING											
Vinyl	C	C	C	CT	CT	C	C	NR	NR	NR	NR
Chlorinated Rubber	C	C	C	CT	CT	NR	C	NR	NR	NR	NR
Bituminous	NR	NR	NR	CT	CT	NR	NR	NR	NR	NR	NR
CHEMICALLY REACTING											
Epoxy	NR	NR	NR	NR	NR	CT	NR	CT	CT	CT	CT
Coal Tar Epoxy	NR	NR	NR	NR	NR	NR	NR	CT	CT	NR	NR
Zinc Rich Epoxy	NR	NR	NR	NR	NR	CT	CT	CT	NR	P	NR
Urethane	NR	NR	NR	NR	NR	NR	NR	NR	NR	CT	C
Polyesters	NR	NR	NR	NR	NR	NR	NR	CT	NR	CT	CT
Inorganic Zinc	NR	NR	NR	CT	NR	CT	NR	CT	CT	P	NR

C = Normally compatible.
CT = Compatible when special surface preparation and/or application conditions are met.
NR = Not recommended because of known or suspected problems.
P = A urethane may be used as a topcoat if its coreactant is of the polyether or acrylic type, but not if it is of the polyester type.

Section 19

Methods for Measuring Coating Properties

Reference: "Paint-Testing Manual," G. G. Sward, Editor,
STP 500, American Society for Testing Materials

Note: Many different properties are measured dependent
on the end use of the coating. Only a few of the
many properties will be singled out for comment.

Optical Properties

Specular Gloss. ASTM Method D523, Specular Gloss. Use standard instrument and measure reflectance at reflectance angle equal to angle of incidence. Three angles are generally used: 20°, 60°, 85°.

Hiding Power. Put coating over black and white pattern and determine thickness of coating required to obliterate pattern.

Tinting Strength. The power of a pigment to color a standard paint or pigment. Usually measured by mixing standard and paint under study and comparing appearance against standards.

Physical Properties

Density. Usually given in weight per gallon. Use standard cups of known weight.

Settling of Pigment. May be accelerated by centrifugation. Usually immerse pan in liquid and follow change in weight with time as pigment settles. When practical, optical density or X-ray density measurements may be made at different levels as a function of time.

Viscosity. Very large number of methods used. These include capillary viscometers, time of flow using other geometries, rotational viscometers, falling ball methods, rising bubble method, band viscometers in which a band is pulled through a block containing the material under study, flowmeters, brushability, sagging, leveling.

Surface Energy. Direct measure of surface tension, contact angle of fluid on surface.

Film Thickness

Wet Film Thickness is measured by a special film gauge or needle micrometer.

Dry Film Thickness is measured by micrometer, needle pentrant gauge, cross sectioning and measuring under the microscope, commercial gauges based on magnetic property, inductance property, eddy current all of which decrease with increase in distance.

Drying Time. Eight stages of drying are recognized:

Set-to-touch

Dust-free

Tack-free (surface dry)

Dry (dry to touch)

Dry-hard

Dry-through (dry to handle)

Dry-to-recoat

Dry print-free

These are measured by techniques which characterize the hardness or tackiness of the surface. The most common instrument is known as the Gardner Circular Drying Time Recorder which presses a stylus against the surface under a standard force. The stylus rotates at a fixed rate and the length of the track is used to characterize the drying time.

Hardness.

Scratch hardness with standard pencils

Damping of pendulum rocker (Sward Rocker)

Indentation

Abrasion Resistance. Falling sand. Abrasion resistance is expressed in terms of volume of abrasive required to wear through a unit thickness of the coating when the abrasive falls from a specified height through a guide tube.

Rate of loss of gloss on mild abrasion.

Rotation Wheel Abrasion - Tabor Instrument. Abrasion is measured in terms of weight loss on a certain number of revolutions under a standard bearing weight.

Rain or water erosion is sometimes measured.

Wet abrasion method. Back and forth motion of a weighted block in the presence of a liquid and an abrasive.

Adhesion.

> Knife test - not quantitative
>
> Chisel test.
>
> Microknife test. The closeness with which parallel lines can be cut in coating without the coating deadhering from substrate.
>
> Force required to strip a unit of organic coating from the substrate using a chisel-like knife.
>
> Crosscut adhesion test. Make a series of parallel cuts in one direction and another set at 90°. The number of squares remaining intact is a measure of the adhesion.
>
> Tensile measurement of coating using an adhesive to bond to the coating.
>
> Lap joint method.

Flexibility. Bending over mandrels of different sizes.

T-bend. The painted metal is bent about 145° and then compressed by a machine so that two ends are parallel. The operation is repeated so as to give 0T, 1T, 2T, 3T, etc. Results are reported as passing the smallest T-bend on which there are no fractures in the coating as observed with a 10X microscope.

Impact tests. Steel ball is allowed to fall from a specific height and the character of the coating is determined either on the impact side or the reverse side.

Tensile strength. Used on free films with standard tensile equipment in miniature form.

Permeability. Things usually measured are water and oxygen. Require free film. Measure rate, by various techniques, of permeation of vapor through the film.

Chemical Resistance.

> Staining
>
> Immerse in water.
>
> Perspiration. Usually coating is held by garter against the calf of the leg and worn.
>
> Exposure to various chemical solutions.

Fire Retardance. Various standard tests are used. UL uses a tunnel. Coating is exposed to gas burners and the amount of smoke and temperature is measured at other end of tunnel with standard draw of air through the tunnel.

Heat Resistance.

Biological Deterioration.

Natural Weathering. Use standard test racks.

Section 20

Corrosion of Painted Metals

Corrosion of Painted Metals—A Review *

HENRY LEIDHEISER, JR. *

Abstract

Seven types of corrosion and precorrosion of painted metals are reviewed: blistering, early rusting, flash rusting, anodic undermining, filiform corrosion, cathodic delamination, and wet adhesion. The importance of the nature of the interfacial region between the coating and the metal is emphasized. Ten unsolved problems are described.

Introduction

Corrosion research has gone through various trends and fashions during the past several decades. The 1950's were marked by emphasis on polarization curves and their applications and new materials of interest to the nuclear energy field, the 1960's by emphasis on stress corrosion cracking and mechanics of metal fracture, and the 1970's by an emphasis on corrosion in unusual environments. The indications are that the major emphasis in corrosion research during the 1980's will be in the area of protective coatings. The severe demands on maintaining integrity of semiconductor devices during service and the fabricating of devices using photosensitive organic coatings are intensifying the interest of semiconductor device manufacturers in protective coatings. Protective coatings include metallic coatings, glass and ceramic coatings, semiconductor coatings, and organic polymeric coatings. It is this latter subject area that this article addresses. The purpose of this presentation is to summarize the present state of knowledge about Corrosion Control by Organic Coatings from the perspective of the author's experience and interests.

The tendency of a coated metal to corrode is a function of three major factors: (1) the nature of the substrate metal, (2) the character of the interfacial region between the coating and the substrate, and (3) the nature of the coating. In many cases, little or no attention is paid to the substrate and the coating is required to compensate for the inadequacies of surface preparation prior to painting. Commercially coated metals, such as the case with automobiles, appliances, and coil coated products, are handled with close attention to every step from the raw material to the finished product. Thus, there are corrosion problems with painted metals that range from improper application procedures to those which are not correctable because of an absence of understanding. It is the intent of this article to focus attention on research aimed at understanding and to stress those aspects where the understanding is inadequate.

An organic coating protects a metal substrate from corroding primarily by two mechanisms: (1) serving as a barrier for the reactants, water, oxygen, and ions, and (2) serving as a reservoir for corrosion inhibitors that assist the surface in resisting attack. The barrier properties of the coating are improved by increased thickness, by the presence of pigments and fillers that increase the diffusion path for water and oxygen, and by the ability to resist degradation. The common degradation mechanisms of the coating include abrasion and impact, cracking or crazing at low or high temperature, bond breakage within the polymer matrix because of hydrolysis reactions, oxidation, or ultraviolet light, and freeze-thaw cycling. The result of such degradation allows access of reactants to the coating/substrate interface without the necessity of diffusion through the polymer matrix. Much work has been published on the degradation of organic coatings and this subject will not be discussed here because it is outside the main thrust of this review.

Sources of Information About the Corrosion of Painted Metals

A novice to the field of coatings science must seek guidance in the library. The important journals in the field that contain articles on corrosion include:

- *Journal of Coatings Technology*
- *Journal of the Oil Colour Chemists Association*
- *Farbe und Lack*
- *Progress in Organic Coatings*

Information on a broad range of coatings types and coatings related issues may be found in a series of pamphlets published by the Federation of Societies for Paint Technology in Philadelphia. These may be purchased collectively in a two-volume set. An excellent summary of paint defects and suggestions for correction may be found in "Hess's Paint Film Defects—Their Causes and Cure."[1] A well organized summary of the properties and resistance to chemical attack of a wide variety of paints may be found in the book, "Design and Corrosion Control."[2] Some outstanding articles on various types of protective coatings will be found in the Encyclopedia of Materials Science and Technology to be published by Pergamon Press in 1983. Two books which include papers given at international symposia are very useful: "Corrosion Control by Coatings"[3] and "Corrosion Control by Organic Coatings" published by NACE.[4]

Types of Corrosion Beneath Organic Coatings on Metals

This review will focus on six specific types of corrosion beneath organic coatings: blistering, early rusting, flash rusting, anodic undermining, filiform corrosion, and cathodic delamination. Each of these will be treated separately. Another type of coating deterioration known as "loss of adhesion when wet," or simply "wet adhesion," which may or may not be related to corrosion, will also be discussed.

---

* This is the first in a series "Current Topics in Corrosion" which are invited reviews by the editor. Submitted for publication March, 1982.

* Center for Surface and Coatings Research, Lehigh University, Bethlehem, Pennsylvania.

Blistering

An up-to-date review of blistering has been given recently by Funke.[5] Blistering is one of the first signs of the breakdown in the protective nature of the coating. The blisters are local regions where the coating has lost adherence from the substrate and where water may accumulate and corrosion may begin. Five mechanisms, operative under different circumstances, are used to explain blister formation which occurs prior to the corrosion process.

Blistering by Volume Expansion Due to Swelling.[6-9] All organic coatings absorb water and those used in corrosion protection are usually in the range of 0.1 to 3% water absorption upon exposure to liquid water or an aqueous electrolyte. Water absorption leads to swelling of the coating and when this occurs locally for any reason, blisters may form and water may collect at the interface.

Blistering Due to Gas Inclusion or Gas Formation.[10] Air bubbles or volatile components of the coating may become incorporated in the film during film formation and leave a void. Such blisters are not necessarily confined to the interface, but when they are, they can serve as a corrosion precursor site.

Electroosmotic Blistering.[11-14] Water may move through a membrane or capillary system under the influence of a potential gradient. Potential gradients, such as may exist with a galvanic couple, have the capability of leading to a blister.

Osmotic Blistering.[15,16] The driving force for osmotic blistering is the presence at the coating/substrate interface of a soluble salt. As water penetrates the coating to the interface, a concentrated solution is developed with sufficient osmotic force to drive water from the coating surface to the interface and a blister is formed. The osmotic mechanism is probably the most common mechanism by which blisters form.

An outstanding example of osmotic blistering was cited by an unidentified discussion participant at the Corrosion 81 meeting in Toronto. A ship was painted in Denmark and made a voyage immediately thereafter across the Atlantic and into the Great Lakes. When it reached port, a blister pattern in the form of a handprint was observed above the water line. Apparently, the paint was applied over a handprint. No blistering occurred during exposure to sea water because of the high salt content of the water, but when the ship was exposed to fresh water, the osmotic forces became significant and the blistering occurred.

Blistering Due to Phase Separation During Film Formation.[17-20] A special type of osmotic blistering can occur when the formulation includes two solvents, the more slowly evaporating one of which is hydrophilic in nature. When the hydrophilic solvent is in low concentration, the phase separation process occurs at a later stage in film formation and may occur at the coating/substrate interface. Water diffuses into the hydrophilic solvent, or into the void left by the hydrophilic solvent, and blisters are initiated. Glycol ethers or esters, which have low volatility, are prone to cause such blister formation.

In all the above cases, the blister provides a locale for collection of water at the coating/substrate interface. Oxygen penetrates through the coating, leaching of ionic materials from the interface or from the coating occurs, and all the constituents are available for electrochemical corrosion. The rate of reaction appears to be controlled by the oxygen permeability of the coating.[19,20] Oxygen is necessary for the cathodic reaction, $H_2O + 1/2O_2 + 2e^- = 2OH^-$, but it is also consumed in the conversion of Fe(II) to Fe(III). The ferric corrosion products tend to concentrate on the inside dome of the blister and at the periphery of the blister where the oxygen concentration is the highest. The cathodic region is at the periphery of the blister and the anodic region is in the center of the blister where the oxygen concentration is the lowest.

Early Rusting[21]

This term is applied to a measles-like rusting that occurs after the coating has dried to the touch. It only occurs after the coated metal is exposed to high moisture conditions. A typical condition under which it is observed has been cited by Grourke.[21] A steel tank was abrasively cleaned and was then painted with an acrylic latex late in the afternoon during the summer. Rust spots were most prominent on the bottom of the horizontally-mounted tank and were observed up to the liquid level within the tank. The top half of the tank exhibited no rust spots.

The three conditions which lead to early rusting are: (1) a thin latex coating (less than 40 μm; (2) a cool substrate temperature; and (3) high moisture conditions. Early rusting can be duplicated in the laboratory under these conditions. The severity of the problem tends to increase as the activity of the steel is increased. For example, early rusting is more severe on panels given a white abrasive blast[22] in comparison with less adequate cleaning using a power tool.

Early rusting occurs with latex coatings because of the mechanism by which they lose water. Film formation occurs through coalescence of the latex particles.[23,24] Particle-particle contact occurs because of water evaporation, particle-particle deformation then occurs as a consequence of surface tension and capillary forces, and finally diffusion of polymer chains occurs among latex particles and the film hardens. Early rusting occurs under those conditions that slow down the rate of drying and allow water soluble iron salts to be leached through the paint film. If moist conditions do not exist during the latex drying process, early rusting does not occur.

In summary, early rusting is a consequence of moist conditions occurring before the latex coating has dried sufficiently. Water ingress and egress occur readily before particle coalescence has been completed and movement of soluble iron salts through the film, followed by water evaporation leads to the rust staining. Success in preventing early rusting has been obtained through the use of soluble inhibitors in the formulation.

Flash Rusting[25]

Brownish rust stains may appear on blast-cleaned steel shortly after priming with a water based primer. This phenomenon is known as "flash rusting." Work done by the Paint Research Association[26] has shown that this defect may be avoided by removing the contaminants remaining after blast cleaning either by careful cleaning or by a chemical treatment before application of the primer.

The steel or ceramic grit on the surface apparently leaves crevices and/or galvanic cells are set up between the steel grit and the steel base sufficient to activate the corrosion process as soon as the surface is wetted by the water based paint. The staining is a result of the soluble corrosion products penetrating the coating and being oxidized to the ferric form within or on the surface of the coating.

The possible adverse effect of steel grit blasting on the performance of paints is a subject worthy of investigation.

Anodic Undermining

Figure 2 shows six planes along which delamination may occur to separate the organic coating from the metal. Anodic undermining represents that class of corrosion reactions underneath an organic coating in which the major separation process is the anodic corrosion reaction under the coating. An outstanding example is the dissolution of the thin tin coating between the organic coating and the steel substrate in a food container. In such circumstances, the cathodic reaction may involve a component in the foodstuff or a defect in the tin coating may expose iron which then serves as the cathode. The tin is selectively dissolved and the coating separates from the metal and loses its protective character.

Aluminum is particularly susceptible to anodic undermining. Koehler[27] cites an example from laboratory studies in which pairs of organic coated aluminum panels were sealed to opposite ends of a cylindrical cell filled with 0.05 citrate solution at pH 3.5 containing 0.5% NaCl. The two panels were connected to a power source and a current of 0.09 $\mu amp/cm^2$ was

FIGURE 1 — Typical corrosion that was preceded by cathodic delamination. This type of corrosion is common on painted steel which has been damaged by impact or by abrasion. Corrosion is visible as brown blisters along edge to right of gas tank cover and along chrome trim of window at top of picture.

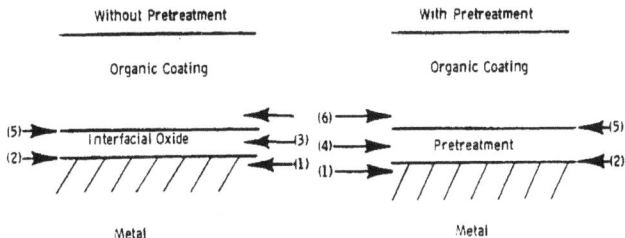

FIGURE 2 — Schematic representation of the six planes along which delamination may occur in an organic coating/metal substrate system. Note that relative thicknesses of oxide and pretreatment layers and the organic coating are not to scale.

passed for one week. Underfilm corrosion was observed on the anode and no underfilm damage occurred on the cathode.

Anodic undermining of organic coatings on steel may occur under circumstances where the steel is made anodic by means of an applied potential. In the absence of an applied potential, coatings on steel fail largely by cathodic delamination.

Anodic undermining has not been studied as extensively as cathodic delamination because there do not appear to be any mysteries. Galvanic effects and principles which apply to crevice corrosion provide a suitable explanation for observed cases of anodic undermining.

Filiform Corrosion

Filiform corrosion is a type of attack in which the corrosion process manifests itself as thread-like filaments. It represents a specialized form of anodic undermining. It generally occurs in humid environments and is most common under organic coatings on steel, aluminum, magnesium, and zinc (galvanized steel). In some cases, filiform corrosion will develop on uncoated steels on which small amounts of contaminating salts have been accidentally deposited. It has also been observed on thin electrodeposits of tin, silver, gold, and under conversion coatings such as phosphate. The most annoying types of filiform corrosion are those which occur under paint films that are designed to retain an aesthetic appearance on metals exposed to the atmosphere and those which occur under the lacquers which protect the interior of food containers.

The threads which form in filiform corrosion exhibit a wide variety of appearances from nodular shapes such as those on aluminum to the very fine, sharply-defined threads observed under clear lacquers on steel. The widths of the filaments are of the order of 0.05 to 0.5 mm and under laboratory conditions grow at the rate of 0.01 to 1 mm/day. The rate of growth of the filaments is approximately constant over long periods of time.[28,29] Filiform corrosion requires a relatively high humidity, generally over 55% at room temperature, but insufficient work has been done to determine an exact lower limit. It can be encouraged to develop on scratching through the coating to the metallic substrate and then maintaining the panel at a relative humidity of 70 to 85%. It develops in the laboratory in some cases simply by putting powders of a salt on the surface of the coating or by putting salt crystals on the

metal substrate and applying the coating over the contaminated surface. Preston and Sanyal,[29] for example, obtained filiform corrosion on steels at 99% relative humidity by inoculating the steels with the following powders before applying the coating: sodium chloride, calcium sulfate, ammonium sulfate, sodium nitrate, zinc chromate, flue dust, cinders, iron oxide, and carborundum.

Hoch[30] has made a detailed study of the character of filiform corrosion on steel, magnesium and aluminum substrates. In the case of iron, the very leading edge of the filament had a pH of approximately 1, whereas the liquid immediately behind the leading edge had a pH of 3 to 4, just what would be expected on the basis of hydrolysis of Fe^{++} ions. Magnesium and aluminum also exhibited very low pH's at the leading edge and higher pH values in the liquid adjoining the leading edge. Insoluble corrosion products form in all three cases a short distance back from the leading edge. In the cases of aluminum and magnesium, hydrogen bubbles were observed at the leading edge, indicating that the liquid in the leading edge was low in oxygen and the hydrogen evolution reaction took precedence.

The following mechanism appears to account satisfactorily for the filiform corrosion of aluminum coated with an organic lacquer. First, a highly localized defect forms in the coating. This defect may arise at the edge of a scratch through the coating, at an inclusion in the coating, at a defect in the metal surface, or as a result of the presence of a local high concentration of electrolyte which causes penetration of the coating. In the presence of high relative humidity, water penetrates through the coating. In the presence of electrolyte, which has either penetrated the coating or was inadvertently occluded beneath the coating, a tiny liquid aggregate forms because of the high affinity of ions such as Na^+ and Cl^- for water. Once sufficient molecules are present to have a liquid-like identity, additional water diffusing through the coating is retained because of the low vapor pressure of concentrated electrolyte solutions. Minor corrosion of the substrate occurs yielding additional dissolved ions and promoting further retention of diffusing water species. As the liquid increases in dimension and corrosion occurs, local conditions cause an imbalance in the oxygen supply at some point in the microscopically circular corrosion area. The oxygen-deficient area becomes the anode and the periphery becomes the cathode. The circular droplet then assumes an elliptical shape and the conditions for filamentary growth are present. Once the filament has been nucleated, there is developed an oxygen concentration cell and the propagation of the filament proceeds because of a highly effective anode at the head and a cathodic area present in the areas surrounding the head. Immediately at the interface of the growing head, aluminum is dissolved to yield a highly concentrated Al^{+++} ion solution. Hydrolysis oc-

3

curs with the following reactions liberating H^+ and thus generating the very low pH:

$$Al(H_2O)_6{}^{+++} = Al(H_2O)_5OH^{++} + H^+$$

$$Al(H_2O)_5OH^{++} = Al(H_2O)_4(OH)_2{}^+ + H^+$$

$$Al(H_2O)_4(OH)_2{}^+ = Al(H_2O)_3(OH)_3 + H^+$$

The latter reaction which occurs some distance from the leading edge results in the precipitation of hydrated aluminum oxide and the pH is reduced to the range of 3 to 4 because of dilution from incoming water.

The oxygen deficiency at the very leading edge along with the low pH also permits the competing cathodic reaction, $2H^+ + 2e^- = H_2$, to occur to a limited extent and a small amount of hydrogen gas is generated.

Koehler[31] has emphasized the importance of the anion in filiform corrosion. He noted that the filaments formed on steel in the presence of sulfate contaminants beneath the coating were less numerous and were much finer than those observed with chloride contaminants. He also noted that the head of the filament contained the anion of the contaminating salt, but not the cation. The anions apparently migrated to provide charge compensation for the ferrous ions formed in the active region at the head of the filament.

The fascinating question about filiform corrosion is why the corrosion occurs in the form of filaments as opposed to circular spots. No complete answer is possible at the present time but it does appear that the limited availability of oxygen, by diffusion through the coating, and the limited availability of water, by diffusion through the coating under relative high humidity conditions, are the determining factors. At very high relative humidities or on exposure to liquid water, filiform corrosion passes over to more general corrosion and the filamentary character is lost.

Much attention has been paid in paint laboratories to reducing filiform corrosion. Phosphate conversion coatings, followed by chromate rinses and distilled water rinses, provide some protection, but do not completely eliminate, the filiform corrosion of iron. The properties of the coating also have an effect on the extent and character of filiform corrosion. Coatings that are highly permeable to water and to oxygen are specially susceptible to filiform attack. Coatings that are very brittle and are ruptured by pressures generated by the corrosion process lose the entrapped moisture and pitting attack often result. In the case of magnesium, filiform corrosion occasionally converts to a virulent pitting attack. More information can be found in reference 17.

Cathodic Delamination

Many coated steel products are subject to scratches or dents with consequent exposure of the steel to the environment. If the coated materials are continuously immersed in an electrolyte, as for example, ships, underground pipelines, and the interior of vessels holding an aqueous solution, it is possible to protect the exposed areas by an applied cathodic potential. One of the undesirable consequences of cathodic protection is that the coating adjoining the defect may separate from the substrate metal. This loss of adhesion is known as "cathodic delamination." This type of delamination may also occur in the absence of an applied potential. The separation of the anodic and cathodic corrosion half reactions under the coating provides regions which are subject to the same driving force as when the cathodic potential is applied externally.

It is generally believed[33,34] that the major driving force for cathodic delamination in corrosion processes in the presence of air is the cathodic reaction, $H_2O + 1/2O_2 + 2e^- = 2OH^-$. When an applied potential is used, the important reaction may be $2H^+ + 2e^- = H_2$, if the driving force is sufficient. Figure 3 shows a typical cathodic polarization curve for steel in 0.5M NaCl saturated with air. The regions of dominance of the two

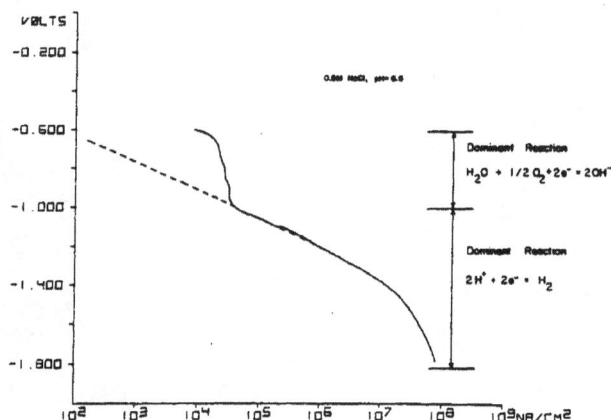

FIGURE 3 — Potentiodynamic polarization curve for steel in aerated 0.5M NaCl. Dotted line represents the extrapolation of the portion of the polarization curve that represents the hydrogen evolution reaction.

cathodic reactions are noted on the figure and the cathodic polarization curve for hydrogen evolution in the absence of oxygen is shown by the dotted line. It is apparent from this figure that at a potential of -0.8 V (vs SCE), the dominant reaction is hydrogen evolution. Polarization at -0.8 V of polymer-coated steel containing a defect in the absence of oxygen leads to no significant delamination from the defect, whereas in the presence of air there is significant delamination.

Studies indicate that the pH beneath the organic coating where the cathodic reaction occurs is highly alkaline, as the cathodic equations indicate. Ritter and Kruger[35] have recently reported that the pH at the delaminating edge is greater than 14 as measured by pH-sensitive electrodes inserted through the metal substrate from the back side. Other studies which integrate the pH over a larger volume of liquid beneath the coating yield pH values of 10 to 12. Cathodic polarization curves on steel in 0.5M NaCl at pH values of 6.5, 10, and 12.5 are approximately the same[36] suggesting that the cathodic behavior beneath the coating may be rationalized in terms of the cathodic polarization curve that is applicable at the exposed defect.

Figure 4 represents the extremes of three types of polarization curves that are observed on metals whose behavior during cathodic treatment has been studied. Point A on each curve represents the potential at the defect and Point B represents the assumed potential at the delaminating front. Curve (1) has the shape of the polarization curve of aluminum in 0.5M NaCl. The surface is not active for the oxygen reduction reaction and the rate of delamination is low as indicated by the location of Point B. Curve (2) is a hypothetical curve somewhat comparable to the polarization curve for tin in 0.5M NaCl. The curve has a steep slope such that the current density falls off greatly with increase in potential. Thus the current density at the delaminating front is low. Curve (3) is typical of iron and copper in 0.5M NaCl. The oxygen reduction reaction is catalyzed over a wide potential range and the current density remains the same over this range. Cathodic delamination occurs at a relatively rapid rate because the current density at Point B is high relative to comparable points on curves (1) and (2).

The cathodic reaction which occurs at the delaminating front generates hydroxyl ions which appear to be the major destructive influence on the organic coating/substrate bond. The value of the pH at the delaminating front is determined by the following factors: the rate at which the reaction occurs; the shape of the delaminating front; the rate of diffusion of hydroxyl ions away from the delaminating front; and buffering

4

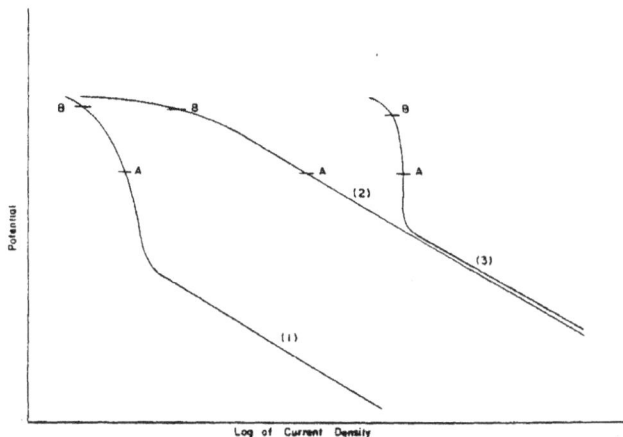

FIGURE 4 — Three types of cathodic polarization curves observed on different metals in aerated 0.5M NaCl. Curve (1) is obtained with aluminum; curve (2) is hypothetical and illustrates the approximate curve obtained with tin; and curve (3) is obtained with steel.

reactions which may involve the interfacial oxide or the polymer.

All the evidence presently available indicates that the cathodic delamination process occurs because of the high pH generated by the cathodic reaction. The real question is what is the consequence of a high pH on the interface. The experimental evidence suggests that the strong alkaline environment may attack the oxide at the interface or may attack the polymer. Attack of the oxide has been seen by Ritter[37] using ellipsometric techniques in the case of polybutadiene coatings on steel and surface analysis techniques in the hands of Dickie and colleagues[38] give clear evidence that carboxylated species are present at the interface as a result of hydroxyl ion attack of the polymer. Dickie[39] has also recently shown that coatings more resistant to alkaline attack exhibited better performance when scribed and submitted to salt spray.

It is proposed that the major mechanism for the delamination process is the solubilization of the thin oxide coating at the interface between the organic coating and the metal. The dissolution of the oxide, as the major delaminating mechanism, has been proposed previously by Gonzalez, Josephic, and Oriani[40] to account for effects observed with food can lacquers on steel when immersed in solutions containing strong complexing agents for Fe^{++} ions. In this process the oxide itself participates in the cathodic reaction by the reaction,

$$\gamma\text{-}Fe_2O_3 + 6H^+ + 2e^- = 2Fe^{++} + H_2O ,$$

and the complexing agent serves to drive the reaction to the right by complexation of the ferrous ions. In the cathodic delamination process being proposed herein, the dissolution of the oxide breaks the bond between the coating and the substrate metal and the high pH leads to localized attack of the polymer at the interface. The presence of oxidized organic species on the metal surface after delamination has occurred may be the result of *a posteriori* adsorption of oxidized species or may be the result of islands of organic left on the surface. The XPS technique used in reference 38 illuminates a large area and spatial resolution is lacking to determine if the organic material is present over the entire surface or is island-like in nature.

Since the important delaminating process is a consequence of the hydroxyl ions generated by the cathodic reaction, $H_2O + 1/2O_2 + 2e^- = 2OH^-$, the delamination may be prevented by any of the following:

> Preventing reactant water from reaching the reaction site.
> Preventing reactant oxygen from reaching the reaction site.
> Preventing electrons from reaching the reaction site.
> Preventing cation counterions from reaching the reaction site.
> Reducing the catalytic activity of the surface for the cathodic reaction.

Water and oxygen reach the reaction site largely by diffusion through the coating[41] so the rate of reaction may be reduced by reducing the permeability of the coating for these constituents. It appears unlikely on the basis of present knowledge to eliminate completely the diffusion of these constituents through the coating or through defects in the coating that are present when the coating is prepared or occur during normal service. The electrons, however, reach the reaction site through the metal phase and through any interfacial oxide or other film that exists at the metal/coating interface. Any type of metal surface film that is a poor electronic conductor has the possibility of limiting the access of electrons to the reaction site. The low rate of the cathodic delamination from aluminum surfaces, relative to zinc and steel, is probably a consequence of the poor electronic conductivity of the aluminum oxide at the interface.

The cathodic reaction generates anions and there must be available cation counterions to balance the charge locally. Hydrogen ions do not perform this function or else the pH would not rise as dramatically as it does. Available evidence[41] indicates that the major transport medium for the cations is in the liquid layer that forms at the coating/substrate interface, although transport through the coating itself cannot be ruled out.

The cathodic reaction is a catalyzed one and the chemical character of the interfacial region determines whether or not the reaction will occur. It has been shown that cobalt ions have the ability to poison the zinc oxide surface on zinc for the oxygen reduction reaction[42] and it has also been shown that this poisoning leads to a lower rate of cathodic delamination.[43] Some commercial pretreatments possibly include the function of poisoning the surface for the cathodic reaction.

Loss of Adhesion of a Coating When Wet. This phenomenon shows up in a number of ways. The best example is exhibited by the "hot water test" used in characterizing pipeline coatings. A candidate coating on a steel substrate is immersed in water at 80 to 100 C for 10 to 30 days and the adherence of the coating is determined at the end of this time by the use of a strong and sharp knife. A similar test of a duration of approximately 1 hour is used with certain coil coated products. Poor wet adhesion is also exhibited by coatings that dry too rapidly and trap organic solvents at or near the interfacial region. This phenomenon has been discussed by Funke.[44] It also is apparent in the removal of epoxy powder coatings using methyl ethyl ketone. After soaking the coated metal in methyl ethyl ketone for one or more days, the coating may be readily stripped from the substrate if the time after removal from the organic liquid is correctly selected. In too short a time, the coating is gummy and cannot be separated from the substrate. In too long a time, adherence to the substrate is regained. When the correct time is chosen, the coating may be removed in much the same way as a weakly adhering adhesive tape. At the correct time, there apparently is a layer of liquid which is intermediate between the coating and the substrate along with the fact that the coating has a physical state intermediate between a soft clay and a rigid coating.

The major problem in studying wet adhesion is the difficulty in quantitatively defining the degree of adherence. The cross-cut test is used with some degree of success with coatings less than about 100 μm in thickness but is not suitable for very thick coatings. The impact test of Zorll[45] is useful with coatings in the same thickness range as the cross-cut test. Unfortunately, there is no satisfactory test for coatings 100 μm or greater in thickness. A vertical tear off test, commonly used for coating adherence tests, is not applicable to thick, inflexible coatings but it can be used to determine wet adhesion of thin coatings in some cases. Funke[44] has proposed a test which as yet has not received wide acceptance. He suggests that the water absorption of both free films and of the same material on a metal be determined at 90% relative humidity. Loss of adhesion on exposure to high humidity is indicated by a cross-over in the two absorption curves. The cross-over time indicates when the coated metal begins to lose adhesion because of the presence of water at the metal/coating interface. Coatings which exhibit good wet adhesion do not show a cross-over.

The major unsolved problem relating to wet adhesion is the mechanism by which water affects the adhesion. This problem is related to the very basic question as to the factors that control adhesion. The way in which adhesion between two materials is viewed has gone through a number of fashions. Some years ago, adhesion was viewed as an interaction between polar groups; later it became the rule to treat adhesion in terms of dispersion forces. More recently, under the impetus of Fowkes,[46] adhesion of organic coatings is being viewed as an acid/base interaction. This approach has much merit because the acid/base character of a surface can be quantitatively assessed and the adsorption of organic polymers on metals and metal oxides can be studied to determine the appropriateness of this latter viewpoint. The effect of water on the rate and amount of polymer adhesion shows promise of providing the basis for understanding wet adhesion.

Nature of the Interfacial Region

Much research has been carried out with the objective of characterizing the metal surface before the application of the coating. Some recent examples include the work of Schwab and Drisko[47] on the effect of surface profile, the work of Mansfeld, Lumsden, Jeanjaquet, and Tsai on surface chemistry,[48] the work of Zurilla and Hospadaruk[49] on characterizing the oxygen reduction capacity of phosphated surfaces, and the work of Iezzi and Leidheiser[50] on the effect of parameters which control the ability of a steel surface to accept uniform phosphating.

The chemical nature of the intact organic coating/metal substrate interfacial region has been little studied largely because of the difficulty in devising experimental techniques to make such a study. Optical techniques, such as ellipsometry, are useful with thin transparent coatings but they cannot be used with opaque coatings or those which contain pigments or fillers. Ellipsometric techniques can provide information on the thickness of the oxide film at the interface and, with good fortune, the optical parameters may be used to determine the composition of the oxide. In the hands of Ritter and Kruger,[51] it has been found that an oxide exists at the interface and that the optical properties of the interface change when cathodic delamination or corrosion occurs.

Emission Mössbauer spectroscopy provides a technique which may be utilized to identify the chemical nature of the emitting atom. Chemical compounds yield characteristic spectra which can be used as fingerprints to identify the compound. Leidheiser, Kellerman, and Simmons[52] have applied Mössbauer techniques to show chemical changes beneath a coating and, more recently, Leidheiser, Simmons, and Music[53] have been able to show changes in the nature of the oxide film on cobalt when a polybutadiene coating was applied. The simple application of the coating caused a fractional conversion of Co^{+++} to Co^{++} and baking the coating at 200 C resulted in the formation of additional Co^{++}. The emission Mössbauer technique has the severe limitation that it is only applicable with relative ease to cobalt and tin and the complete interpretation of the spectrum is often difficult and/or ambiguous.

Major advances in organic coatings protective against corrosion are dependent on the development of new techniques for studying in a nondestructive way the chemical nature of the interfacial region between the metal and the organic coating. The interfacial region is where the action is, particularly in the case of cathodic delamination. The oxide at the interface is the catalytic surface for the oxygen reduction reaction; it is the medium through which the electrons are supplied for the oxygen reduction reaction; and it provides the bonding which results in the adherence between the coating and the metal.

Commercial systems which provide the maximum resistance against corrosion include an inorganic coating between the organic coating and the metal substrate. This inorganic coating, often called a "pretreatment" or a conversion coating, replaces the normal metal oxide and provides to the organic coating a substrate with different chemical properties—a poorer catalyst for the oxygen reduction reaction, a less conductive interfacial region and, in some cases, a rougher interface that improves organic coating/substrate adherence and resistance to deterioration under service conditions. The more common inorganic coatings include phosphates, chromates and mixed metal oxides. With the exception of the research of Machu,[54] little has been published on the science associated with interfacial inorganic coatings. It is a fertile area for research.

Properties of the Coating

General Comments

Polymers form the matrix of organic coatings. They are the retainers for pigments, fillers, corrosion inhibitors, and other additives present for specific purposes. The polymer selected is based both on end-use requirements and the ability to apply the coating in the desired manner. Emulsion polymers, or latexes, are suspended in an aqueous medium and they form a coating by loss of water by evaporation and coalescence of the individual particles into a continuous film. Some monomers, such as polybutadiene, are applied to the substrate in a solvent and the polymerization process occurs as the solvent is removed. Cross-linking by oxidation occurs at elevated temperature in the case of polybutadiene. Other coatings based on condensation polymers are polymerized *in situ*. A good example is the epoxy-polyamine coating in which the two constituents are mixed just prior to application and the polymerization process occurs over a period of time. Other polymers are dissolved in solvents and the polymer forms the coating as the solvent is evaporated. Production line painting often involves the use of heat or other type of radiation to cause the film-forming process to occur more rapidly.

The properties of polymeric coatings depend not only on the size, shape and chain structure of the individual units, but also on the spatial shape of the polymer molecules. The linking of many carbon atoms and the freedom of rotation about carbon-carbon bonds permits the molecule to assume a variety of spatial shapes such as spirals, coils, and tangles. This wide latitude in shape also leads to a variety of ways in which individual molecules are oriented with respect to their neighbors. Three classes of arrangement are recognized:

1. Segments of the molecule are randomly distributed regardless of whether they belong to the same molecular chain or another chain. Such a material is termed amorphous or glassy and the properties are uniform in all directions.

2. Segments of the molecule possess a degree of lateral order through the folding of individual chains. The volume element over which this occurs may be considered a single crystal. The individual crystals may be randomly oriented or they may be aligned in the same direction. In the latter case, the coating may have physical properties that differ in different directions.

3. Segments of the molecule may show lateral order through the parallel arrangement of extended chains. As in (2),

these parallel arrangements may be unoriented with respect to neighboring volumes or there may be a degree of spatial orientation. Materials of this type are obtained when a polymer melt is solidified while under shear or stress.

An unusual case of corrosion in which the rate of corrosion appears to be related to segmental motion of portions of the polymer chain is shown in some interesting research by Yializis, Cichanowski, and Shaw.[55] These workers observed that the rate of corrosion of aluminum coated polypropylene capacitors in either the dry state or when immersed in a dielectric fluid was a function of the frequency of an applied AC potential. A sharp maximum in corrosion rate occurred at 3.5×10^3 Hz. The following explanation for this phenomenon is offered. On exposure to the atmosphere and during corona discharge prior to metallization, polypropylene dissolves significant quantities of oxygen and water. Appreciable amounts of chloride ion remain in the polypropylene from the manufacturing operation. When the metallized capacitor is exposed to an AC voltage, segmental motions occur in the polypropylene. Over a limited frequency range, the motions of segments of the polymer are such as to allow diffusion of water and oxygen ot occur along special pathways in the polymer. Since the aluminum coating is essentially opaque to the passage of water, chloride ions and oxygen, sufficient reactants accumulate at the aluminum/polymer interface to allow the following reaction to occur:

$$\text{Cathodic: } H_2O + 1/2 O_2 + 2e^- = 2OH^-$$

$$\text{Anodic: } Al - 3e^- = Al^{+++}$$

The chloride ion provides sufficient conductivity in the aqueous phase to allow the electrochemical reactions to occur and prevents the formation of a passive film of aluminum oxide at the aluminum/polypropylene interface.

An important characteristic of polymers and organic coatings is known as the glass transition temperature, T_g. It is that temperature at which a discontinuity occurs in a physical property as a function of temperature when the polymer exists in the amorphous condition. Typical physical properties which show discontinuities at T_g include the coefficient of expansion and specific heat. It is interpreted as that temperature above which the polymer has sufficient thermal energy for isomeric rotational motion or for significant torsional oscillation to occur about most of the bonds in the main chain which are capable of such motion. Values of T_g are obtained by many different experimental techniques, the more common of which include dilatometry, dielectric measurements, spectroscopy, calorimetry, and refractive index. Standish and Leidheiser[56] have outlined a simple technique for such determination of T_g of coatings on a metal using dielectric measurements as a function of temperature. The value of T_g is important because physical properties such as water and oxygen permeability and ductility differ above and below T_g.

The mechanical properties of coatings which are not highly loaded with pigments or fillers depend on molecular weight, crystallinity and the three-dimensional arrangement of the branches. An increase in molecular weight makes a polymer harder and stronger. The higher the degree of crystallinity, the stronger the polymer. Chain polymers containing two different groups, R and R', have different mechanical properties dependent upon the arrangement of the branches. Curing or cross-linking causes a polymer to become harder, more brittle, and less soluble.

Polymer coatings are exposed to the environment and thus are subject to degradation by environmental constituents. The main agencies by which degradation occurs are thermal, mechanical, radiant, and chemical. Polymers may also be degraded by living organisms such as mildew. Deterioration takes the form of discoloration, cracking and crazing, loss of adherence to the substrate, or change in a physical property such as resistivity or mechanical strength.

The mode of degradation may involve depolymerization, generally caused by heating, splitting out of constituents in the polymer, chain scission, cross linking, oxidation, and hydrolysis. Polymers are subject to cracking on the application of a tensile force, particularly when exposed to certain liquid environments. This phenomenon is known as environmental stress cracking or stress corrosion cracking and there are many analogies to similar phenomena observed with metals.

Zinc-Rich and Zinc-Pigmented Coatings

The term "zinc-rich" applies to coatings which contain up to about 95% metallic zinc in the dry film. They are electrically conductive and protect the metal substrate electrochemically in much the same way as zinc protects steel in galvanized steel. The vehicle is usually a silicate in the case of the so-called "inorganic zincs" and various organic polymers are used to obtain so-called "organic zincs."

Zinc-pigmented paints are usually a mixture of metallic zinc (80%) and zinc oxide (20%), the latter of which is added to reduce the rate of settling of zinc both in the container and after application to a surface. Galvanic protection does not occur in the case of the zinc dust-zinc oxide paints, whose films do not exhibit electrical conductivity.[57]

The galvanic action of the zinc protects the steel at holidays, cut edges, and scribe marks and a similar galvanic action probably applies during the first stages of corrosion beneath the coating. The major action of the zinc, however, appears to be the sealing of the paint film so that it has an improved resistance to penetration by active environmental species. There is evidence also that the zinc pigment and the zinc oxide prevent deterioration of the inorganic and organic binders and assist in maintaining flexibility and desirable mechanical properties of the coating.

Corrosion Inhibitors

It is thought that corrosion inhibitors in an organic coating function in an identical way to those added to a liquid environment. Corrosion inhibitors used in coatings include oxidizing agents such as chromate, inorganic salts that function in the same manner as benzoates, metallic cations of which lead is the most widely used, and organic compounds. All organic coatings are permeable to water and water contents for many coatings at 100% relative humidity are in the range of 0.1 to 3%. Thus, when the coating is wet, some fraction of the inhibitor is solubilized and can be transported to the metal surface. The coating simply serves as a reservoir for the inhibitor.

A recent article has summarized six technical requirements for an ideal corrosion inhibitor to be used in organic coatings.[58] These include the following:

1. The inhibitor must be effective at pH's in the range of 4 to 10 and ideally in the range of 2 to 12.

2. The inhibitor should react with the metal surface such that a product is formed which has a much lower solubility than the unreacted inhibitor.

3. The inhibitor should have a low but sufficient solubility.

4. The inhibitor should form a film at the coating/substrate interface that does not reduce the adhesion of the coating.

5. The inhibitor must be effective both as an anodic and a cathodic inhibitor.

6. The inhibitor should be effective against the two important cathodic reactions, $H_2O + 1/2 O_2 + 2e^- = 2OH^-$ and $2H^+ + 2e^- = H_2$.

A critical problem facing the coatings industry at the present time is the need to replace chromates and lead compounds as corrosion inhibitors. Both of these classes of materials have been identified as hazardous and the pressures to remove them from formulations are growing. Adequate proven substitutes have not yet been universally accepted so that there is an increased interest in developing accelerated tests that can be used to select corrosion inhibitors.

Methods for Monitoring the Corrosion of Polymer-Coated Metals

Electrical methods for studying the protective properties of coatings are numerous and many have produced important results. Two reviews of this subject have been published recently[59,60] and selected information will be extracted from these reviews and other published sources. Electrical measurements that provide data which are useful in predicting the lifetime of a coating include DC measurements of coating conductivity;[62,63] of impedance as a function of frequency;[64,65] of equivalent AC resistance at constant frequency;[66] and of the ratio of capacitive to resistive components at constant frequency.[67] The AC properties of a coating have also been used to estimate the amount of water taken up by a coating.[66,68-74] Scanning techniques[75] have proven useful in characterizing the electrical homogeneity of coatings. The rate of diffusion of sodium chloride through coatings has been determined by Kittelberger[76,77] using measured values of the DC resistance and the membrane potential of the film. Radiotracer measurements,[78] however, are much preferred for these types of measurements. Dielectric techniques are also useful in determining the glass transition temperature of coatings, in determining the effects of coating composition and structure, and in the quality control of coating components.[52]

Corrosion potential measurements and their applicability to coated metals have been summarized by Wolstenholme.[79] As a generalization, it can be concluded that movement of the corrosion potential in the noble direction is indicative of an increasing cathodic/anodic surface area ratio and is indicative that oxygen and water are penetrating the coating and arriving at the metal/coating interface. Movement of the corrosion potential in the active direction is indicative that the anodic/cathodic surface area ratio is increasing and that the overall corrosion rate is becoming significant. Increasingly positive potentials with time suggest that alkaline conditions caused by the oxygen reduction reaction are developing locally at the metal/coating interface and that delamination is of concern. Increasingly active potentials indicate rusting (in the case of steel) beneath the coating and represent the signal that the coating lifetime is limited.

Important Unsolved Problems

The unsolved problems related to corrosion and its prevention by organic coatings are legion. Ten of these problems which are considered of special importance are mentioned and briefly discussed. Three of these problems relate to the very practical test development and corrosion monitoring, two relate to a better understanding of practical operations, and five relate to a better understanding of the coating/substrate interface.

1. **The Development of an Accelerated Atmospheric Corrosion Test, the Results of Which Correlate Well with Service Experience.** The most commonly used test for evaluating the corrosion protection by a coating is the salt spray test, used either with an undamaged coating or used with a coating that is scribed in the form of an X. Many workers are dissatisfied with the salt spray test because there is often a lack of correlation with service experience. However, it is a test that is widely used and it is the most likely test to be acceptable both to the supplier of a coating and the potential user of a coating. Other tests involve simulated outdoor exposure with the degradation variables of ultraviolet radiation, temperature, humidity, and cycling taken into account.

It is the author's opinion that the most satisfactory type of accelerated test will be one that provides information about the extent of reaction at the coating/substrate interface on a localized scale and at a time long in advance of visible corrosion on the surface of the coating. Such a test should be capable of being used on test panels of different shapes and upon exposure to different aggressive environments.

2. **The Development of a Satisfactory Test for Screening Inhibitors for Use in Coatings.** Environmental considerations are limiting the use of two satisfactory inhibitors, lead oxides and salts, and chromates. It can be anticipated that other inhibitors will be recognized as harmful and will be unavailable for use by coatings fomulators. No satisfactory test for rapidly screening inhibitors in a coating formulation has yet received wide acceptance. Inhibitors that perform satisfactorily while dissolved in the aggressive medium do not necessarily perform satisfactorily when incorporated in the coating.

3. **The Development of a Corrosion Monitor That Gives Information About the Progress of Corrosion and Allows One to Predict the Lifetime Before Repainting is Necessary.** Many coatings are applied in locations that are not easily accessible. Examples include towers, bridge spans, submerged supports for deep sea platforms, underground pipelines, and hidden parts of metallic structures. It would be very useful to have a cheap device that would periodically sense the coating/substrate system, and transmit the information to a central location so that the system could be followed as a function of time. The most likely type of device would be one that applied an AC potential at a fixed frequency, or perhaps a range of frequencies, and converted this information to capacitive and resistive components. It is difficult to visualize a test that does not depend on an electrical measurement, although long range inspection by fiber optics may be acceptable.

4. **What is the Mechanism of Corrosion or Adhesion Loss of Coated Metals in the Hot Water Test?** Coatings for certain applications are sometimes tested for adhesion by exposing the coating to water at 80 to 100 C for an hour or more and the adhesion is very poor when tested immediately after removal from the hot water, although the coating recovers good adhesion when it is permitted to dry out. What is the mechanism for the loss of adhesion when wet? And why is the adhesion regained when the coating has dried?

Underground pipeline coatings are often subjected to a hot water test for periods of time up to 20 days. Some coatings survive this test and, indeed, show no deterioration of the coating/substrate bond. Other coatings show severe corrosion that appears to correlate with corrosion potential measurements made continuously during the hot water test. These measurements suggest that some coatings are defective at the start of the test. There is need to understand the mechanism of this behavior.

5. **What is the Reason for the Poor Performance of Some Coatings When the Coatings are Applied Over Steel Substrates That are Abrasively Cleaned with Steel Grit?** As stated previously, flash rusting is associated with contaminants remaining on the surface after abrasive cleaning. Work in this laboratory has shown that certain coatings applied over a steel grit blasted surface exhibit poorer performance during exposure to hot water than similar coatings prepared over surfaces blasted with aluminum oxide. Much steel grit remains imbedded in the surface after cleaning as shown by the fact that the steel is released by superficial oxidation at 300 C. Research is needed to determine the essential reasons for the adverse effect of steel grit on corrosion performance.

6. **Development of a Better Understanding Why Cathodically Electrodeposited Organic Coatings Provide Such Good Protection Against Delamination in Chloride-Containing Environments.** Cathodic electrocoats, or E-coats as they are often called, have received wide acceptance in the automobile industry because of the good protection they offer against the delamination process that accompanies corrosion around a break in the coating. There appears to be little understanding of the mechanism by which these coatings operate. One possibility appears to be that the strongly reducing environment at the steel surface coincident with the formation of the coating/steel bond leads to an interface with good integrity. Experiments which seek to characterize the nature of the steel surface during electrocoating, after electrocoating, after baking, and after exposure to an aggressive medium are planned utilizing Mössbauer emission spectroscopy.

7. Development of a Method for the Detection of Condensed Water at the Coating/Substrate Interface. Corrosion beneath an organic coating will occur only if there is an aqueous phase to accept the cations formed in the corrosion process and to provide the medium in which the oxygen reduction reaction may occur. Thus, any method for detecting the presence of aqueous-phase water at the interface and the increase in volume of water at the interface as a function of time will have the capability of detecting corrosion at the earliest possible time.

The question is simple but the answer is difficult. Preferably the method should be sensitive to water aggregates of the order of 25 to 100 molecules, but a lower sensitivity would be acceptable. The method should be applicable to opaque coatings and should be able to discriminate between water masses within the coating and condensed water at the interface. Electrical methods or spectroscopic methods to which the coating is not opaque appear to offer the greatest chance of success.

8. Achieving a Better Understanding of the Potential Distribution Within a Delaminating Region. It has been shown previously[36,41] that cathodic delamination may be interpreted in terms of polarization curves. It has also been shown in the case of steel that the polarization curve in 0.5M NaCl solution is not strongly a function of pH over the range of 6 to 12.5. The missing information needed for a complete interpretation of the delamination process is the potential distribution radially from a defect when the metal is polarized cathodically. The mathematical treatment is difficult because the following information is not known: the thickness of the aqueous layer at the interface as a function of distance from the defect; the conductivity of the liquid; the amount of charge that passes through the coating radially from the defect; the oxygen concentration gradient across the coating.

9. Development of Information about Ionic Transport Through the Coating When the Metal Substrate is Polarized Cathodically. The cathodic reaction that occurs under the coating during corrosion or during cathodic polarization generates OH^- ions. These ions require counterions to maintain charge neutrality. Since the pH rises, the counterions cannot be exclusively H^+ but must be largely alkali metal ions such as Na^+. In one set of experiments carried out in this laboratory, a current of the order of 10^{-11} amp/cm^2 developed across a thin organic coating when the coated metal was coupled to an uncoated metal of equal surface area in 0.5M NaCl. The coated metal was the cathode. It would be desirable to know if the Na^+ ions represent the charge carriers across the coating under the experimental conditions used.

Experiments should be carried out to determine the ability of alkali metal ions to migrate through various coatings under a potential gradient and this ability should be compared with the ability of a coating to resist cathodic delamination. It is essential in experiments of this type to use a coated metal, as opposed to a free film, since the boundary conditions are so different in the two cases.

10. Development of a Better Understanding of the Chemical Nature of the Bond Between the Organic Coating and the Substrate Metal and How the Bond Changes with Time. The interfacial region in the absence of a purposely applied chemical conversion coating consists of the metal, a thin oxide coating on the metal, perhaps a water layer, and the organic coating. The bond between the organic coating and the substrate has been variously referred to as a hydrogen bond, a Van der Waals bond, an acid/base interaction, an electrostatic bond, etc. Minimal experimental information is available in specific instances to characterize the bond quantitatively. A subsidiary question is what happens to the nature of the bond when water permeates the coating and becomes available for adsorption or reaction in the interfacial region.

Acknowledgment

This review was written and some of the experimental results reported herein were obtained while the author's research was supported by a major grant from the Office of Naval Research. This support is gratefully acknowledged.

References

1. Hess's Paint Film Defects, Edited and revised by H. R. Hamburg and W. M. Morgans, 3rd Ed., Chapman and Hall, London, 504 pp. (1979).
2. Design and Corrosion Control, V. Roger Pludek, Halsted Press, New York, 383 pp. (1977).
3. Corrosion Control by Organic Coatings, H. Leidheiser, Jr., Editor, Science Press, Princeton, N.J., 500 pp. (1979).
4. Corrosion Control by Organic Coatings, H. Leidheiser, Jr., Editor, Natl. Assoc. Corrosion Engrs., Houston, Texas, 300 pp. (1981).
5. W. Funke. Prog. Organic Coatings, Vol. 9, p. 29 (1981).
6. N. A. Brunt. J. Oil Colour Chem. Assoc., Vol. 47, p. 31 (1964).
7. N. A. Brunt. Verfkroniek, Vol. 33, p. 93 (1960).
8. L. Bierner. Farbe Lack, Vol. 66, p. 686 (1960).
9. D. M. James. J. Oil Colour Chem. Assoc., Vol. 43, p. 391, 658 (1960).
10. J. A. van Laar. Paint Varn. Prod., Vol. 51, No. 8, 31, 88; No. 9, 49; No. 11, 41, 97 (1960).
11. H. Grubitsch and K. Heckel. Farbe Lack, Vol. 66, p. 22 (1960).
12. H. Grubitsch, K. Heckel, and R. Sammer. Farbe Lack, Vol. 69, p. 655 (1963).
13. H. Grubitsch, K. Heckel, and O. Monstad. Farbe Lack, Vol. 70, p. 167 (1964).
14. W. W. Kittelberger and A. C. Elm. Ind. Eng. Chem., Vol. 39, p. 876 (1947).
15. L. A. van der Meer-Lerk and P. M. Heertjes. J. Oil Colour Chem. Assoc., Vol. 58, p. 79 (1975).
16. L. A. van der Meer-Lerk and P. M. Heertjes. J. Oil Colour Chem. Assoc., Vol. 62, p. 256 (1979).
17. W. Funke. J. Oil Colour Chem. Assoc., Vol. 59, p. 398 (1976).
18. C. M. Hansen. Ind. Eng. Chem. Prod. Res. Dev., Vol. 9, p. 282 (1970).
19. W. Funke and H. Haagen. Ind. Eng. Chem. Prod. Res. Dev., Vol. 17, p. 50 (1978).
20. W. Funke, E. Machunsky, and G. Handloser. Farbe Lack, Vol. 84, p. 49 (1978).
21. M. J. Grourke. J. Coatings Technol., Vol. 49, No. 632, p. 69 (1977).
22. According to specifications of Steel Structures Painting Council, Pittsburgh, PA.
23. J. W. Vanderhoff, E. B. Bradford, and W. K. Carrington. J. Polym. Sci., Vol. 41, p. 155 (1973).
24. M. S. El-Aasser and A. A. Robertson. J. Paint Technol., Vol. 47, No. 611, p. 50 (1975).
25. M. J. Grourke and T. H. Haag. Resin Rev., Vol. 24, No. 1 (1974).
26. Paint Research Assoc. Newsletter No. 7 (1978).
27. E. L. Koehler. Localized Corrosion. R. W. Staehle. B. F. Brown, J. Kruger, and A. Agrawal. Editors, Natl. Assoc. Corrosion Engrs., Houston, Texas. p. 117 (1974).
28. H. Kaesche. Werkstoffe Korros., Vol. 10, p. 668 (1959).
29. R. St. J. Preston and B. Sanyal. J. Appl. Chem., Vol. 6. p. 26 (1956).
30. G. M. Hoch. Localized Corrosion. R. W. Staehle. B. F. Brown, J. Kruger, and A. Agrawal, Editors, Natl. Assoc. Corrosion Engrs., Houston, Texas. p. 134 (1974).
31. E. L. Koehler. Corrosion, Vol. 33, p. 209 (1977).
32. ASTM Standard G8-72, Annual Book of ASTM Standards, Vol. 27, p. 869 (1979).
33. H. Leidheiser, Jr. and M. W. Kendig. Corrosion, Vol. 32, p. 69 (1976).
34. R. A. Dickie and A. G. Smith. Chem. Tech. 1980, No. 1, p. 31.
35. J. J. Ritter and J. Kruger. Corrosion Control by Organic Coatings, H. Leidheiser, Jr., Editor, Natl. Assoc. Corrosion Engrs., Houston, Texas. p. 28 (1981).

36. H. Leidheiser, Jr., L. Igetoft, W. Wang, and K. Weber. Proc. 7th Intern. Conf. on Organic Coatings, Athens, Greece (1981); in press.
37. J. J. Ritter. Personal communication, October 1981.
38. J. S. Hammond, J. W. Holubka, and R. A. Dickie. J. Coatings Technol., Vol. 51, No. 655, p. 45 (1979).
39. R. A. Dickie. paper presented at Electrochem. Soc. meeting, Minneapolis, Minn., May 1981.
40. O. D. Gonzalez, P. H. Josephic, and R. A. Oriani. J. Electrochem. Soc., Vol. 121, p. 29 (1974).
41. H. Leidheiser, Jr., W. Wang, and L. Igetoft. Prog. Organic Coatings, accepted for publication.
42. H. Leidheiser, Jr. and I. Suzuki. J. Electrochem. Soc., Vol. 128, p. 242 (1981).
43. H. Leidheiser, Jr. and W. Wang. Corrosion Control by Organic Coatings, H. Leidheiser, Jr., Editor, Natl. Assoc. Corrosion Engrs., p. 70 (1981).
44. W. Funke. Corrosion Control by Coatings, H. Leidheiser, Jr., Editor, Science Press, Princeton, N. J., p. 35 (1979).
45. U. Zorll. Adhasion, Vol. 6, p. 165 (1979).
46. F. M. Fowkes, C-Y. Chen, and S. T. Joslin. Corrosion Control by Organic Coatings, H. Leidheiser, Jr., Editor, Natl. Assoc. Corrosion Engrs., Houston, Texas, p. 1 (1981).
47. L. K. Schwab and R. W. Drisko. Corrosion Control by Organic Coatings, H. Leidheiser, Jr., Editor, Natl. Assoc. Corrosion Engrs., Houston, Texas, p. 222 (1981).
48. F. Mansfeld, J. B. Lumsden, S. L. Jeanjaquet, and S. Tsai. Corrosion Control by Organic Coatings, H. Leidheiser, Jr., Editor, Natl. Assoc. Corrosion Engrs., Houston, Texas, p. 227 (1981).
49. R. W. Zurilla and V. Hospadaruk. Soc. Automotive Engrs. Trans., Vol. 87, p. 762 (1978).
50. R. A. Iezzi and H. Leidheiser, Jr. Corrosion, Vol. 37, p. 28 (1981).
51. J. J. Ritter and J. Kruger. Surface Science, Vol. 96, p. 364 (1980).
52. H. Leidheiser, Jr., G. W. Simmons, E. Kellerman. J. Electrochem. Soc., Vol. 120, p. 1516 (1973).
53. H. Leidheiser, Jr., G. W. Simmons, S. Musić. manuscript in preparation.
54. W. Machu. Werkstoffe Korrosion, Vol. 14, p. 566 (1963).
55. A. Yializis, S. W. Cichanowski, and D. G. Shaw. paper presented at IEEE Meeting in Boston, May 1980.
56. J. V. Standish and H. Leidheiser, Jr. J. Coatings Technol., Vol. 53, No. 678, p. 53 (1981).
57. B. C. Hafford. Zinc Dust Metal Protective Coatings, to appear in Encyclopedia of Materials Science and Engineering, M. Bever, Editor, Pergamon Press. Scheduled for publication in 1983.
58. H. Leidheiser, Jr. J. Coatings Technol., Vol. 53, No. 678, p. 29 (1981).
59. H. Leidheiser, Jr. Prog. Organic Coatings, Vol. 7, p. 79 (1979).
60. Y. Sato. Prog. Organic Coatings, Vol. 9, p. 85 (1981).
61. R. C. Bacon, J. J. Smith, and F. M. Rugg. Ind. Eng. Chem., Vol. 40, p. 161 (1948).
62. E. M. Kinsella and J. E. O. Mayne. Br. Polym. J., Vol. 1, p. 173 (1969).
63. J. E. O. Mayne and D. J. Mills. J. Oil Colour Chem. Assoc., Vol. 58, p. 155 (1975).
64. G. Menges and W. Schneider. Kunststofftechnik, Vol. 12, No. 10, p. 265; No. 11, p. 316; No. 12, p. 343 (1973).
65. H. Leidheiser, Jr. and M. W. Kendig. Corrosion, Vol. 32, p. 69 (1976).
66. R. E. Touhsaent and H. Leidheiser, Jr. Corrosion, Vol. 28, p. 435 (1972).
67. J. M. Parks, M. C. Hughes, and H. Leidheiser, Jr. paper presented at Electrochem. Soc. Meeting, Denver, Colo., October, 15, 1981.
68. D. M. Brasher and A. H. Kingsbury. J. Appl. Chem., Vol. 4, p. 62 (1954).
69. C. P. De and V. M. Kelkar. First Intern. Congr. on Metallic Corrosion, Butterworths, London, England, p. 533 (1962).
70. D. M. Brasher and T. J. Nurse. J. Appl. Chem., Vol. 9, p. 96 (1959).
71. J. K. Gentles. J. Oil Colour Chem. Assocn., Vol. 46, p. 850 (1963).
72. R. N. Miller. Materials Protection, Vol. 7, No. 11, p. 35 (1968).
73. H. C. O'Brien. Ind. Eng. Chem., Vol. 58, No. 6, p. 45 (1966).
74. K. A. Holtzman. J. Paint Technol., Vol. 43, No. 554, p. 47 (1971).
75. J. V. Standish and H. Leidheiser, Jr. Corrosion, Vol. 36, p. 390 (1980).
76. W. W. Kittelberger. J. Phys. Colloid Chem., Vol. 53, p. 392 (1949).
77. W. W. Kittelberger and A. C. Elm. Ind. Eng. Chem., Vol. 44, p. 326 (1952).
78. Y. Sato. Denki Kagaku, Vol. 28, p. 538 (1960).
79. J. Wolstenholme. Corrosion Science, Vol. 13, p. 521 (1973).

Section 21

Accelerated Corrosion Tests for Evaluating Protective
Properties of Organic Coatings

Two important references:

(1) ASTM Book of Standards, Part 27, "Paint Tests for Formulated
Products and Applied Coatings," ASTM, 1916 Race St., Philadelphia,
PA 19103.

(2) Federal Standard 141a, Naval Publications and Forms Center,
5801 Tabor Avenue, Philadelphia, PA 19120.

Six Tests will be described very briefly.

(1) Water Immersion. The coated metal is immersed in deionized water
for 24 hours or more at temperatures from room temperature to 100°C. The
test conditions are a function of the type of coating being tested. Blisters
in the coating or lowered adhesion (poor wet adhesion) indicate the following:
(a) excessive salt or water soluble residues under the coating; (b) contamination of the paint coating with soluble salts; (c) improper and inadequate
pretreatment.

(2) Filiform Corrosion. Clear lacquers, varnishes and paints with low
pigment to vehicle concentration are prone to a special type of corrosion
known as "filiform corrosion." The coated metal is submitted to salt spray,
rinsed with distilled or deionized water, and placed while still wet in an
exposure cabinet or a closed system in which the temperature is maintained
at 25°C and the relative humidity at 85%. Specimens are examined weekly for
six weeks for evidence of the thread-like tunnels characteristic of filiform
corrosion.

(3) Salt Spray Test. The coated metal is scribed with a sharp tool in
the shape of an "X". The metal is exposed to salt spray and the degree of
undercutting of the coating adjoining the scribe marks is measured. The less
the undercutting, the more resistant the coating is to "cathodic delamination."

(4) Cycle Tests. Many different cycle tests are being studied in
commercial laboratories but none has yet been standardized and accepted by
ASTM. One of the tests advocated by General Motors is the following:

Day	Duration	Exposure
Monday	24 hours	Salt spray, ASTM B117
Tuesday	8	Humidity Cabinet (100°F, 100% RH)
Wednesday & Thursday	16	Humidity cabinet, air and humidity shut off, lid of cabinet closed
Friday	8	Humidity cabinet (100°F, 100% RH)
Friday, Saturday, Sunday	64	Humidity cabinet off and cabinet open

This cycle is repeated up to 10 times.

(5) Arizona Proving Ground Scab Corrosion Cycle. Panels are scribed through to the substrate and are subjected to 60 24-hour cycles of the following sequence:

(a) Immersion in 5% NaCl solution, air saturated, for 15 minutes.

(b) Exposed at room temperature with no rinsing or wiping for 1¼ hours.

(c) Humidity cabinet exposure at 120°F and 85-90% RH for 22.5 hours.

Scab corrosion is reported in millimeters total width of scab.

(6) Cathodic Delamination Test. Used for coatings that will be subjected to cathodic delamination. A hole is drilled through the coating and the coated metal, with edges protected from electrolyte, is immersed in 3% NaCl solution. The metal is polarized to -1.5 volt vs SCE and is held constant at this potential by means of a potentiostat. At the end of a fixed time, which may be as much as 30 days, the extent of delamination of the coating radially from the exposed metal is determined.

There are many other tests that are designed for specific purposes which include such factors as sunlight, condensing humidity, thermal gradients, temperature cycling, environmental constituents such as sulfur dioxide, etc.

Section 22

Removal of Organic Coatings

Note: This section has been taken from "Technology of Paints,
        Varnishes and Lacquers," Edited by C. R. Martens.

As paint formulations improve, it becomes more of a problem to remove coatings in order to apply a new coat of paint. Special stripping compositions have been developed to keep up with advances in modern paint technology.

Coatings which are particularly difficult to remove are catalyzed epoxies, polyurethanes and thermosetting acrylics. The older the coating, the more difficult it is to strip.

There are two broad types of paint removers: the solvent type and the chemical type which is usually alkaline in nature.

Paint Removal

Paint and varnish removers which are usually applied by brush but can be applied by spray or other methods must contain a thickener so that the paint and varnish remover will stay on a vertical surface, such as a wall, long enough to soften the paint. For coatings that are difficult to remove, the painted part is immersed in a tank of the stripper.

The problem of removing a coating is often very complex. Some of the factors which influence the problem of a stripping operation are:

   (1) Type of film former.

   (2) Thickness of coating.

   (3) Type of surface applied.

   (4) Primer used.

   (5) Type of pigment.

   (6) Curing time, temperature and method.

   (7) Age of coating.

Formulation

The correct paint and varnish remover for a given application is the one which does its task satisfactorily at the lowest cost. An inexpensive material which requires the use of heat, is slow acting, or takes several applications may cost more than a higher-priced remover.

When choosing a paint remover, the nature of both the coatings and the substrates must be taken into consideration. Frequently a single stripper must work on a variety of materials, whose composition is often unknown. In this case, methylene chloride strippers are generally used. However, when a paint remover is needed for a specific application such as stripping of rejects of a coating of known composition, then several considerations should be made.

Most common substrates are unaffected by solvent-based strippers. Some exceptions are plastics and certain aluminum alloys which are corroded slightly by chlorinated hydrocarbons or their decomposition products. Some of the modifiers incorporated into organic solvents may have undesirable effects. For example, amines may discolor wood or react with copper or cadmium surfaces, and acids may corrode ferrous and magnesium alloys. Aqueous strippers should not be used on wood since water raises the grain. Caustic rapidly attacks aluminum and zinc alloys.

Solvent Paint and Varnish Removers

The most popular type of paint and varnish remover is the solvent type. These removers contain a primary solvent plus a co-solvent, activator, thickener, evaporation retarder, corrosion inhibitor and emulsifier (for water rinsable type).

A number of factors must be considered in formulating a solvent-type remover. All of the desirable features may not be realized in one product, but the goals are rapidly of action, correct viscosity, nontoxicity, nonflammability, low odor, clean rinsability, noncorrosiveness and package stability.

Solvents. Since the active ingredient is the solvent, most attention is paid to its selection. By far the most popular paint remover solvent is methylene chloride. It is practically nonflammable, thus minimizing fire hazards, one of the greatest dangers in paint removal. Methylene chloride is one of the least toxic of the chlorinated hydrocarbons. It is also one of the most efficient paint removing solvents.

In general, the efficiency of a chlorinated solvent decreases as the chlorination or the chain length of the organic radical increases.

One method of rating the efficiency of a solvent for paint removal is the time required to wrinkle a standard (oleoresinous) paint film. As methylene chloride is the most effective, it is given the top rating as shown below.

Methylene chloride	100
Chloroform	69
Ethylene dichloride	45
Trichloroethylene	36
Monochlorobenzene	36
Carbon tetrachloride	24
o-Chlorobenzene	22
Propylene dichloride	15
Trichlorobenzene	10

Other solvents which will soften paint films are (in approximate decreasing order of efficiency): ketones, esters, aromatic hydrocarbons, alcohols and aliphatic hydrocarbons. Many less common solvents are useful for specific applications, but they are too costly for widespread use. Some of these are: 2-nitropropane, dimethylformamide, dimethyl sulfoxide, tetrahydrofuran and 1,1,2-trimethoxyethane.

Co-solvent. Combined with methylene chloride, methanol is one of the less expensive and most frequently used co-solvents. The general rule to follow in adding co-solvents to methylene chloride is that polar solvents reduce stripping time, nonpolar solvents do not. Small concentrations of polar solvents are more effective than large concentrations except when used on air-dried phenolic coatings and shellac.

Activators. Activators enhance penetration of the solvent into the protective coating. With a methylene chloride remover, the addition of enough water to saturate the methylene chloride will reduce the time as much as 90% for some formulations. A small amount of acetic acid (0.2%) will speed up removal further.

Amines and other alkaline ingredients are often used as activators. Some of these are ammonia, monoethylamine, morpholine, etc. Amines, however, sometimes react with chlorinated solvents and stain wood.

Thickeners. The thickener must meet four major qualifications: (1) It must impart high viscosity at low solids concentration. (2) It must be compatible with the blended composition. (3) It should maintain uniform viscosity during storage. (4) It should form a soft, nonadhering film upon drying. The thickened system should be thixotropic for easy application and nonsag properties on vertical surfaces.

A number of substances, organic and inorganic, are used as thickeners. The most important are organics such as methylcellulose, ethylcellulose, cellulose acetate and nitrocellulose. Also used as thickeners are materials such as bentonite, metallic soaps, starch, zein, casein, polyacrylate esters, etc. If surface-active agents are used, methylcellulose is practical in a flush-off formula. The viscosity of the thickener is influenced not only by the solvent but also by the co-solvent. The thickener in flush-off formulations should be water soluble or at least water dispersible in the presence of emulsifying agents.

Penetrants. Wetting agents, primarily amines, not only promote wetting power, but also augment the penetration of the coating and make rinsing easier. Other agents used are petroleum sulfonates and ethylene glycol monobutyl ether.

Evaporation Retarders. Evaporation retarders in paint and varnish removers are waxes of petroleum origin. Upon application the wax crystallizes out to form a film on the surface, retarding the evaporation of the solvent. The concentration of wax should be kept at a minimum, as all wax must be removed from the surface before repainting since wax will soften, decrease adhesion and possibly hinder the drying of the new paint.

A typical formulation is shown in Table 22-1.

Table 22-1

Paint and Varnish Removers
(Wash-off Type)

Methylene chloride	71	gal
Toluene	3	"
Methanol	12	"
Di-triisopropanolamine	9.5	"
Water	1.5	"
Methylcellulose (4000 cps)	13	lb
"Areskap"[a]	33	"
Potassium oleate	22	"
Paraffin (122-124°F, mp)	16.5	"

[a]Registered trademark, Monsanto Co.

Chemical Removers

A boiling solution of caustic soda, at a concentration of a few pounds per gallon, is an effective and inexpensive paint remover, and is often used for general stripping purposes. Additives such as sequestering agents, surfactants and activators are added to the caustic to increase stripping rates. Gluconic acid and alkali metal gluconates are good sequestering agents in highly alkaline solutions. Sequestering agents aid in the removal of paints containing oxide pigments and keep metallic ions from precipitating in alkaline solution. The function of the surfactant in these removers is to act as a wetting agent and to emulsify and remove the partially decomposed paint film from the surface. Surfactants that are stable in hot alkalies are sodium resinate, fatty acid soaps, sodium lignosulfonate and petroleum sulfonates. Activators such as phenols and their sodium salts are added to speed removal.

Table 22-2 shows a formula for a dry alkaline paint stripping compound.

Table 22-2

Dry Alkaline Paint Stripper

	Weight (lb)
Sodium hydroxide	85
Sodium lignosulfonate	6
Sodium gluconate	5
Cresylic acid	3
"Nacconol"[a]	1
	100

[a]Registered trademark, Allied Chemical
   Corp., National Aniline Division.

Nonchlorinated Solvent Paint Removers

Because of their low cost, strippers based on solvents other than chlorinated hydrocarbons are still used for removing oleoresinous finishes and less
resistant finishes. They are of the lacquer solvent type containing aromatic
solvents such as benzene and toluene with co-solvents such as methanol and
sometimes ketone solvents. Activators, thickeners and waxes of the same kind
used with methylene chloride can be incorporated.

Phenols and chloroacetic acids are highly active chemicals which are useful in specific situations. Specific materials which are commercially available at low cost are cresylic acid (mixed cresols) and a crude mixture of
mono-, di- and trichloracetic acids.

Removal of Epoxies. For the removal of epoxy coatings, acidic-type materials are particularly effective. One of the most effective strippers for
this purpose is a solution of concentrated nitric acid in dimethyl sulfoxide
used in the temperature range of 120 to 130°C.

Mechanism of Paint Removal

In order to understand why a particular paint remover is more effective
at stripping certain types of coatings than others, it is helpful to consider
how a paint remover works. Simple dissolution of the film is almost never
observed except when shellac or lacquers are removed with certain solvents.
Usually the coating is swelled, softened and raised so that it can be quickly
removed by scraping or flushing with water. In solvent-type removers, the
film is swelled by the solvent. In many types of coatings (e.g. oils or
alkyds), materials such as alkali saponify or break down the ester linkage,

thereby aiding removal. The remover attacks the weakest part of the film and increases penetration.

Methods of Paint Removal. There are four basic methods used in stripping paints. A hot flow-on method is most economical for removing multiple layers of paint from large areas. The remover, usually an alkaline type, is heated to temperatures of 180 to 212°F and applied through a spray head which is a perforated pipe or rake. The runoff is collected and recirculated. A second method is a steam gun, in which an alkaline-type remover is applied at 5 to 10 pounds of pressure.

Tank stripping is applicable where small work must be stripped. This is an excellent method for reclaiming parts in volume. Parts remain immersed until the paint is loosened. Agitation will speed paint removal. For tank immersion stripping, a welded steel tank with an overflow dam to carry away loosened paint is recommended. Where stripping is done with hot solutions, a closed steam coil or immersion-type gas burner is required. Once paint has been loosened, a high-pressure rinse effects complete removal.

Manual brushing is used when dealing with large vertical or inverted surfaces where it is not desiarble to set up a hot flow on a steam gun stripping operation. The paint is removed manually by brushing on a viscous solvent-type stripper. After the paint is softened, it is loosened by steam or water pressure.

Section 23

Hazards in the Preparation and Application of Organic Coatings

Note: This section has been taken from Chapter 8 of "Hess's
       Paint Film Defects"

This section deals exclusively with those hazards which may confront any-
one having to handle paint, particularly when applying it, and also those
hazards associated with its storage. Any reference to the hazards centered
on the actual manufacture of paint has been deliberately omitted as outside the
scope of this book. It aims to provide health and safety information relating
to paint whether the user be a do-it-yourself enthusiast, a professional house
decorator or maintenance man, or the employer or employee of a works manufacturing products such as cars and engineering components where paint is utilized not
only for decorative but for protective purposes.

By "hazard" is meant danger to health or even to life itself. Paint, in
general, presents a two-fold potential hazard: chemical poisoning, and fire.
The accent, however, is on the word "potential". By exercising due care in the
light of what is known of the hazards of a particular paint, no one need ever
be harmed by it whether he be a handyman making spasmodic use of the material or
one who depends for his livelihood on painting and decorating.

Returning now to the toxicity and fire aspects of paint, we are today
witnessing a changing scene. Frequently associated in the past years with the
hazards of poison has been lead, more often in the form of the compound white
lead. The toxic hazards of lead and its compounds are dealt with in more
detail later on, together with the toxicity of other pigments, as well as of
solvents and numerous other components of modern paints. It is conceivable that
the benefits of a decreasing usage of white lead might have been neutralized by
the introduction of numerous other substances, some with properties, hitherto
unencountered, capable of harming mankind. Happily, better education and increasing information on the toxicological properties of paint raw materials
have seen to it that such hazards as are associated with painting are much
better controlled than they were in the 19th century.

A similar situation appertains to the fire hazard. Although the advent
of emulsion finishes and other water-based paints has done much to reduce the
danger of fire, numerous new solvent-borne paints, e.g. the acrylics, have been
coming along to perpetuate the risk. One must not forget, too, that the highly
flammable nitrocellulose finishes are still with us, albeit principally for
industrial usage. Again, one can record with satisfaction that legislation
and authority, information and education, have all played their part, not only
in controlling the fire hazard, but in ensuring that the paint user knows pre-
cisely what the attendant fire risks are and what he has to do to minimize
them.

Toxic Hazards

The word "toxic" covers a multitude of sins. In the extreme case, it is applied to anything which causes fatal poisoning but has also come to mean substances which are merely injurious to health to varying degrees.

Types of Toxicity.

(a) Poisoning

May be sub-divided further into:

   (i) Acute poisoning, i.e., attended with symptoms of severity, as opposed to

  (ii) Chronic poisoning which is recurring or of long continuance.

 (iii) Insidious poisoning, i.e., working secretly, so that the victim is unaware his health is gradually deteriorating and, as is so often the case, is unprepared for remedial action until it is too late.

High concentrations of phenol taken internally or accidentally splashed in large volume on the skin will quickly cause acute poisoning whereas long exposure, say by inhalation, to low concentrations of lead compounds will lead to chronic lead poisoning.

There are three distinct classes of insiduous poison all with what be termed delayed effects. They are the carcinogens, mutagens and teratogens. Chemical carcinogens are substances which induce cancer in man or animals. Less than a century ago there were no recognized chemical carcinogens. Today there are many hundreds of known animal carcinogens; however, only a handful of these are proven human carcinogens. Oddly enough, not every human carcinogen has been demonstrated to be an animal carcinogen, a notable example being arsenic and its compounds.

Vinyl chloride is sometimes met with in the form of its co-polymer with other vinyl compounds. Until recently, vinyl chloride was not implicated as a chemical carcinogen. Now, it is recognized as a causative agent in angiosarcoma, a comparatively rare form of liver cancer, which has been found in a statistically significant number of workers exposed to vinyl chloride monomer during the production of p.v.c.

Again, while the inhalation of asbestos particles has been known as liable to lead to asbestosis, which like silicosis is a crippling lung disease, it is only within recent years that it has been ascertained that exposure to asbestos, particularly the crocidolite mineral, can also lead to malignant mesothelioma, or cancer of the pleura (lining of the lung cavity). Mesothelioma invariably takes a fatal course; fortunately it is a comparatively rare condition. Asbestos powder was at one time a frequent component of various types of industrial paints; now its use as a paint raw material is actively discouraged. It could of course be argued that provided the hazards associated with the handling of asbestos powder during paint manufacture are kept under control, the final product, containing asbestos particles thoroughly dispersed in the paint medium, will present only a minimal health hazard. The health hazard could, however, be regenerated if an asbestos-containing paint film came to be incinerated or even dry-flatted.

Not only carcinogens, but mutagens, too, are of increasing concern, even at the international level. Mutagens are substances which are capable of altering the body's genetic material, i.e., the genes and chromosomes of the cells. Only within the last 15 years has the relationship between man and his environment become a demanding issue. It is environmentally-oriented research which has contributed greatly to the literature on chemical mutagenesis. For instance, by 1974 the Environmental Mutagens Information Centre of the Oak Ridge National Laboratory (U.S.A.) had on file over 10,000 data entries containing information on approximately 4000 different compounds, many of which were environmental pollutants. Are any of such substances met with in paint? No one can give a categorical "no". For example, ethylene oxide, a non-carcinogen, has been found to have mutagenic properties. This mono-epoxide goes into the manufacture of numerous non-ionic surfactants. In addition, claims have been made that ethylene oxide monomer has been detected in streams contaminated by trade effluent, presumably containing minute amounts of unreacted ethylene oxide!

A better understanding of the properties of naturally occurring substances and, more importantly, the rapid growth of the chemical industry have contributed to the development of the new science of teratology. Teratogens are substances which, taken by a mother-to-be, can adversely affect her unborn child. The classic example of an extremely teratogenic substance is the drug thalidomide. The tragic consequences of its introduction subsequently stimulated considerable research activity to the point that monthly publications now exist devoted solely to abstracting the literature on teratology. Again, what has this to do with paint? Around the early 1970s, those industries such as plastics and paints, which made use of plasticizers were disturbed to learn that gross fetal abnormalities resulted when rats were injected with various phthalate esters, notably di-2-ethyl-hexyl phthalate. Top level discussions subsequently pointed to the fact that the experiments involved excessive amounts of injected phthalate and eventually the hue and cry died down early in 1973, when authoritative opinion (including the American F.D.A.) declared that even plastic containers in which phthalate ester was incorporated did not present an indirect food hazard.

What can we learn from all this? The paint user cannot be expected to be fully conversant with carcinogens, mutagens, and teratogens. More and more, with technical information and toxicological data at their call, paint manufacturers are becoming his guardian, protecting him from the potential health hazards associated with the vast array of new chemicals being made available to the paint industry. The corollary to this is that the purchaser of paint is well advised to rely on paint produced by reputable manufacturers, many of whom strive to minimize the hazards of their products, over and above what is required by legislation.

(b) Skin effects

The greatest potential of paint to harm the user probably lies in the adverse effects it can have on the skin. There are three main contributive factors for this:

(i) The heterogeneous composition of paints, containing, as they so often do, skin-degreasing solvents as well as a great variety of pigments and miscellaneous additives, some of which can be physiologically active chemicals.

(ii) The fact that, more often than not, paint involves the use of brush, roller or other hand-operated equipment and, in the greater number of instances, no hand protection is used.

(iii) The regrettable fact that so many painters, even those who operate spray-guns, will often, and against good advice, clean their hands with solvent.

Human skin is a complex membrane, extensive in area, and subject to many diseases. Skin consists of several anatomically recognized layers but there are only two main divisions:

(i) The outer layer, called the epidermis, epithelium or cuticle.

(ii) The true skin or dermis, whose thickness can range from about 3 mm - 0.5 mm.

Throughout the skin are numerous sebaceous (oil- or fat-secreting) glands and sweat glands. The latter secrete watery fluid and terminate visibly as skin pores. In addition, the skin carries hair follicles, which originate deep in the true skin, as well as nerve and muscle fibre.

Skin complaints brought on by contact with paint range from mild skin irritation to extreme forms of dermatitis (inflammation often accompanied by flaking or fissuring). There are two principal causes of dermatitis. First and foremost are the primary irritants of which very many are known. They include not only acids and alkalis, and chemicals such as formaldehyde (a frequent component of urea, melamine and phenolic resins), but most of the common organic solvents. The latter act simply by dissolving out the natural grease of the skin, whereas alkalis dissolve the keratin in skin. Other irritants are sensitizers. Some such as maleic acid (or its anhydride), which is occasionally used in the manufacture of certain classes of synthetic resins, do not necessarily cause skin changes at first contact but the effect is observed on repeated contact a few days later. Sensitizers of a different type are the photo-sensitive agents, i.e., substances whose action is triggered-off by sunlight. They include coal tar and pitch and are also met with among synthetic resins (e.g. p-tertiary butyl phenol/formaldehyde resin) and natural resins (e.g. rosin).

Within the United Kingdom, dermatitis is a prescribed industrial disease and could mean the sufferer being entitled to disability benefit. The

Factory Inspectorate issues valuable booklets on "Industrial Dermatitis" which are periodically updated.

The skin, whether it is abraded, broken or even intact, is also a route for internal poisoning [see (c) Poisoning subcutaneously].

(c) Eye effects

The delicate structure of the eye plus the irreplaceable sense of sight renders eye-care of paramount importance. The commonest eye disorder met with in the handling of paint is undoubtedly eye irritation from volatile paint solvents. Sometimes, however, the irritation is primarily due to the presence of minor amounts of volatile chemicals such as unreacted monomer (e.g. acrylates) or pungent additives (e.g. allyl compounds).

A degradation product frequently arising from thermally disintegrated drying oils is acrolein (acraldehyde). This chemical, as well as others such as the previously mentioned allyl compounds, formaldehyde, etc., have the property of being lachrymatory (i.e., they stimulate the tear ducts, making the eyes "water"). Formaldehyde is also the alleged agent responsible for methylated spirits drinkers becoming blind. Methyl alcohol (methanol) taken internally is oxidized to formaldehyde and routed via the blood stream to the optic nerve which it ultimately destroys. Other recorded bizarre effects on the eye which result from chemicals are double vision from exposure to maleic anhydride and hallucinatory symptoms from acrylamide. If the paint (or unreacted component) is strongly acid or strongly alkaline the term corrosive is applicable, and describes their extremely hazardous nature. It is held that, generally speaking, alkalis are more damaging than acids and with a greater capacity to lead to blindness. The argument here is that strong acids tend to precipitate a protein barrier that prevents further penetration into the tissue, whereas alkalis do not do this but continue to soak into the tissue as long as they are allowed to remain in the eye. Particularly destructive of eye tissue are the class of chemicals known as peroxides which demand great care in their handling, even when they are used in small quantities.

No less harmful to the eyes are solid particles such as paint dust from flatted surfaces, as well as liquid particles from the spray-gun or aerosol cans or from mere splashes. Commonly, this results in conjunctivitis or inflammation of the conjunctiva, the moist red layer or mucous membrane in the eye.

Other potential eye hazards reside in carelessly handled spray guns. High pressure jets impinging on the eye can result in the equivalent of mechanical injury, even leading to permanent blindness.

Finally, mention should be made of a common source of eye disorders, rubbing the eye thoughtlessly with hands contaminated with paint chemicals.

(d) Sensitization

Certain substances, drugs, chemicals and even natural products from the vegetable and animal kingdoms can all induce a special sensitiveness in individuals, meaning the capacity to produce an allergic response. The difference between a reaction caused by an allergy and one by a toxic compound is recognized by the fact that the toxic reaction is dose dependent whereas even minute quantities of allergens can provoke severe symptoms.

Sensitization is most often associated with skin disorders and has already been refered to under Skin effects in (b). However, allergens can also be responsible for many other conditions such as fever, arthritis, asthma, headache and dyspepsia. Singled out here for special mention, as being potentially of significance to users of certain types of paint, are the isocyanates with their multi-hazardous properties. Not only are these chemicals irritant (skin and eye), dermatitic and toxic but their vapor, on inhalation, is capable of leading to severe attacks of asthma. Being readily able to bond to proteins, the isocyanates are potent sensitizers.

While exposure to high concentrations of isocyanates may cause asthmatic symptoms in individuals who have not become allergic to them, exposure to concentrations even lower than 0.2 p.p.m. could produce respiratory distress in sensitized persons. Hence the medical profession's current practice of disqualifying from jobs involving exposure to di-isocyanates such as toluene 2.4 di-isocyanate (TDI), workers with a history of asthma, bronchitis or previous allergy to isocyanate.

Di-isocyanates are an essential component of polyurethanes. Sometimes the latter may contain "free", i.e., unreacted isocyanate, hence introducing a potential sensitization hazard.

Routes Available for Poisoning

Leaving aside the artificial routes of injection which are used chiefly in biological experiments, there are three main pathways whereby poisons can enter the body. These are through the mouth (i.e., by swallowing or ingestion), through the lungs (by inhalation) and through the surface of the skin (percutaneously).

(a) Ingestion

Accidental poisoning, *per os*, is comparatively rare, although there is always the possibility of poisoning from food or cigarettes placed on a dirty workbench or contaminated by unwashed hands which have been in contact with paint containing toxic substances. Would-be suicides seldom use paint although it has been known for prison inmates to imbibe the material deliberately in order to seek hospitalization. More common incidents involve distraught parents who have been responsible for leaving an open can of paint accessible to their young child and, in addition, are unsure whether the child has partaken much, some or none of the contents of the can. Equally important in such an emergency is an estimate of the toxicity of the sample of paint in question. Here, toxicology has gone a long way to help and for many, if not most, of the raw materials which go into paint manufacture, toxicity ratings are now available. On the basis of the ultimate composition (by weight) of

-181-

the paint and having available toxicity ratings for the individual components, some assessment can then be made of the paint's overall potential to harm thereby giving the first-aider and the doctor valuable information.

What is this toxicity rating and how is it arrived at? The one most commonly adopted is denoted as LD_{50} and refers to the dose of a substance which can kill 50% of one class of animal. For toxicity tests selected strains of animals are held in captivity thereby eliminating as many variables as possible, and for a particular series of tests one might take dozens of animals to obtain eventually a single mean lethal dose (LD_{50}). It is expressed as so many grams (g) or milligrams (mg) per kilogram (kg) body weight of animal. Thus it is generally held, although there are many exceptions, that the larger the animal, the more poison it can withstand. Likewise, a two-year old child is much more prone than an adult to fatal poisoning by a given weight of toxic substance or quantity of paint of a poisonous nature. Paint components with LD_{50} acute toxicity values (per kilo body weight) of 1 mg (or less), 100 mg, 1 g and 15 g (or more) would be regarded respectively as highly toxic, moderately toxic, slightly toxic, and relatively innocuous.

(b) Inhalation

The respiratory system is the most significant route for poisoning so far as industry is concerned. Substances which can be inhaled range from gases, and the vapors of volatile liquids and solids, to liquid particles in the form of mists and droplets, and solid particles represented by dusts.

Evolution has seen to it that lungs are adapted to inhale air whose approximate composition is nitrogen 78% and oxygen 21%. Among the minor components of air is carbon dioxide which is present to the extent of 0.03%. Moist air will, of course, contain up to a few percent of water vapor. Inhalation of any gaseous mixture with a composition differing substantially from that of air would be definitely harmful and sometimes fatal. Thus one could not even survive for long in an atmosphere of 100% pure oxygen. Nevertheless, oxygen is vital to the existence of human life. Inhaled oxygen is transported from the lungs to the cellular tissue. This movement of oxygen, as well as the reverse transportation of carbon dioxide, is accomplished largely by means of haemoglobin in the red cells of the blood. The latter represents some 8% of one's body weight and the whole volume of this blood (6-7 litres) passes through the lungs in about one minute.

Poisonous gases and vapors can be irritants, asphyxiants, anaesthetics and systemic poisons. Volatile irritants are capable of preferentially affecting specific areas such as the upper respiratory tract, the lung tissue, or the terminal respiratory passages and air sacs. For example, the upper respiratory tract is sensitive to such substances as ammonia, hydrochloric and chromic acids, and alkaline dusts.

There are different types of asphyxiant gases. Firstly, an excess of an inert gas like nitrogen, hydrogen or carbon dioxide will act merely mechanically, by starving the victim of oxygen. Carbon monoxide is highly toxic because it combines with haemoglobin to form a stable compound, thereby depriving oxygen of its means of transportation. High percentages of carbon monoxide are formed when organic materials, such as wood, paper, rubber, resins, etc. are burnt in an atmosphere deficient in oxygen. This will explain why the death of a significant proportion of victims of conflagrations is due primarily to carbon monoxide poisoning.

Organic nitriles, one such is acrylonitrile, a monomer used in the manufacture of polymers and copolymers on which some paints are formulated, are poisonous in the same way as hydrogen cyanide. That is, they combine with cellular catalysts (enzymes), thereby inhibiting tissue oxidation.

Virtually all nitro- and amino-aromatic compounds, many of which (such as aniline, toluidine, dimethyl aniline and nitrobenzene) have been used in paint products, are characterized by their ability to form methaemoglobin in man. Methaemoglobinaemia results from the oxidation of haemoglobin and leads to such symptoms as cyanosis (blueness about the face and extremities), increase in pulse rate, and dizziness.

Anaesthetic substances, by a depressant action on the central nervous system, affect the supply of blood to the brain. In descending order of activity they include:

(i) Olefinic hydrocarbons, i.e., the alkylenes.

(ii) Ether.

(iii) Paraffinic hydrocarbons, e.g. propane, butane and n-hexane.

(iv) Ketones including acetone.

(v) Aliphatic alcohols such as ethanol and propanol

Air contaminants, which on inhalation will cause systemic poisoning, include:

(i) Most halogenated hydrocarbons, such as carbon tetrachloride (liver and kidney poisons).

(ii) Benzene and various phenols.

(iii) Carbon disulphide and methanol (nerve poisons)

(iv) Various derivatives of toxic elements such as mercury, cadmium, manganese, lead, arsenic, phosphorus, selenium and fluorine.

Whereas gases and vapors have ready access to all parts of the respiratory tract, the anatomy of the latter acquires a greater significance when solid particles are inhaled. Inhalation makes use of a passage way starting with the nose, pharynx and larynx and continuing through to the trachea and bronchus. This leads to branching into bronchioles, further branching into alveolar ducts (5 or 6 per bronchiole) and alveolar sacs (3-6 per duct) and finally to the alveoli which are about 0.1 mm in diameter. Large inhaled particles are trapped in the upper portions of the respiratory tract and are subsequently removed towards the mouth by the wave-like motion of ciliated cells. The latter are particularly numerous in the trachea and larger bronchioles. The lung retains particles which range in diameter from 0.3 to 7 μm, most of them being a little over 1 μm in size.

The development of silicosis is largely attributable to the particle size (approx. 0.6 μm) of silica and not to its sharp-edged physical characteristic. Asbestos, on the other hand, causes asbestosis because of its fibrous nature. Fibres with lengths, ranging from 5 - 100 μm are deposited in the lungs where they are trapped and subsequently encapsulated in protein. Because of the resulting engorgement of the lungs, the heart has more work to do supplying oxygen to the blood.

For many countries the best guide of comparative values of inhalational toxicity is that provided by the American Conference of Governmental Industrial Hygienists' (ACGIH) annual lists of Threshold Limit Values (TLVs). These refer to concentrations of air-borne substances below which nearly all workers may be exposed repeatedly, day after day, without adverse effect. Time-weighted average concentrations permit excursions above this limit, provided that they are compensated by equivalent excursions below the limit during the working day. Exceptions are certain substances against which are quoted "ceiling values" which should never be exceeded. e.g. for formaldehyde a ceiling value of 2 p.p.m. is given.

Published TLV's for contaminants range from 5000 parts per million (p.p.m.), by volume, for carbon dioxide down to no exposure as detected by the most sensitive methods for certain specified human carcinogens. Sometimes the TLV is given on a weight basis as so many mg of particulate per cubic meter of air. In the case of perfect gases the conversion of TLVs from a weight basis to a volume basis is facilitated by the knowledge that 1 g mol of the gas has a volume of 24.45 liters at 25°C and 760 mm Hg pressure.

For Great Britain, H. M. Factory Inspectorate of the Health and Safety Executive reissues the ACGIH list of TLVs but also adds a short list of substances (including asbestos and vinyl chloride) for which different values appertain in the United Kingdom.

Other industrial countries have their own lists and include those of the National Swedish Board of Occupational Safety and Health (Stockholm), the Institute National de Recherche et de Securité (Paris), the Czech TLV Committee and the German Research Association/Commission for Testing Harmful Industrial Substances. Values issued by the U.S.S.R. tend to be lower than those of the U.S.A. and Western Europe because they are based on a different physiological assessment. The World Health Organization, in their report No. 415 (Geneva 1969) has listed Safe Concentration Zones for 24 substances, with a recommendation for their adoption internationally.

(c) Poisoning subcutaneously

Mention has already been made under Skin effects of the ability of certain chemicals to act on the skin or immediately below its surface, but nothing has so far been said of passage through the skin as a route to poisoning, alternative to swallowing and inhalation. Abraded skin, cuts and lesions are all naturally liable to facilitate the absorption of poisons. In varying degrees, intact skin also has this capacity. Perhaps it is not so surprising, from what has already been said of the structure of the skin, to learn that it is penetrable by a great variety of toxic substances, which thus find their way into the blood stream and become systemic poisons.

Poisoning by skin contact has been referred to over the ages. Contaminated clothing, gloves, shirts and other garments, are alluded to in Portigliotti's The Borgias. To the naturally-occurring poisons, particularly during this century, industry has added a considerable number of synthetic chemicals capable of acute or chronic poisoning. Even the insidious action of carcinogens is capable of being initiated through the skin.

Not until as recently as 1968 was the chemical propane sultone found to be a powerful carcinogen. By this time it had become widely used in industry and was becoming recognized as of potential value in the manufacture of paint vehicles. H. Druckrey et al. found that even single doses of propane sultone, given subcutaneously to rats, produced a high yield of sarcomas at the site of injection. A little later, Van Duuren et al. reported tumors in 21 out of 30 mice, at the site of injection with propane sultone. This chemical is also a skin sensitizer and repeated applications to the skin can cause extreme irritation. In other words, propane sultone can be expected to lead to cancer on skin contact, at least with animals and without the artificial aid of injection.

Again, it is known that both the simple chemicals ethylene imine and propylene imine are readily absorbed by the skin. Ulland et al. have pointed out that propylene imine is an important chemical intermediate in the production of coating compositions etc., and that it is closely related to its parent homologue ethylene imine which is known to have carcinogenic, mutagenic and teratogenic activity!

How significant skin absorption is as a route for lead poisoning is the sort of question many who handle paint frequently might ask. The answer is, it all depends which compound of lead is involved. Thus inorganic lead compounds, which in general have poor solubility characteristics, are not readily absorbed by the skin. Certain lead salts, however, such as the acetate and oleate are known to be absorbed, possibly by acid (radical) interchange with the natural oils and fats in the skin. Organo-lead compounds, notably lead tetra ethyl (TEL), the ubiquitous petrol anti-knock, are readily absorbed by the skin, often in toxic amounts. Marchenko and Beilikhis report that this property of TEL applies both to the liquid and the vapor.

Another metallic element of interest in the context of paint, and which has skin absorptive properties, is mercury. Even the metal itself, in a fine state of subdivision, is absorbed by the skin. It was, for instance, formerly employed as an inunction (rubbed into the skin after being pre-mixed with oil or fat) in the treatment of skin diseases. On the other hand, through skin absorption, mercury has given rise to chronic poisoning among industrial workers. Experimenting with rabbits, Schamberg et al. came to the conclusion that the respiratory absorption of mercury was of far less importance than its cutaneous absorption.

Silberberg and others have found evidence for the percutaneous absorption of the inorganic compound mercuric chloride and J. Barnes successfully treated a case of acute mercurial poisoning which followed inunction with an ointment containing 1 part in 3 of mercuric chloride.

Direct contact with organic mercurials can cause severe (second degree) burns and, states Goldwater, can result in absorption of measurable amounts of mercury into the body from underclothing and bed linen. An organic mercury compound sometimes used as a paint fungicide is phenyl mercury acetate. Brana Jovicic describes cases of mercury intoxication, believed due to absorption through the skin, among workers handling a fungicide based on a mixture of phenyl mercury acetate and ethyl mercury chloride. More potent are the highly toxic mercury alkyls but these have no particular relevance to paint.

Whereas inorganic tin compounds are of negligible percutaneous toxicity, a totally different picture emerges with the organo-tins. Zerman et al. observed that four cases of poisoning resembled those from lead tetra-ethyl, the tin compounds being readily absorbed by the skin. Percutaneous poisoning by tetrabutyl tin and related compounds in experiments on rabbits was found by Yanagibashi to be reversible within certain limits upon discontinuation of their administration. Local skin washing with 5% potassium permanganate directly after the application of the organo-tin was effective in cases of subacute poisoning. Barnes and Stoner confirmed that many alkyl tins are toxic to animals but drew attention to the fact that di-octyl and di-nonyl derivatives are relatively harmless when applied to the skin in large doses. In experiments with tributyl tin oxide (TBTO), the organo-tin compound which has been used as stabilizer for polyvinyl chloride, as a paint fungicide, and as an active component of anti-fouling paints, Elsea and Paynter found that repeated dermal application to rabbits resulted in no evidence of sensitization. They concluded, however, that because TBTO is a primary skin irritant and can be absorbed through the skin, contact should be avoided. In the event of exposure, contaminated skin areas and clothing should be thoroughly cleaned with soap and water.

Understandably, because of their generally greater lipid solubility, organic compounds have a more enhanced capacity for skin absorption than inorganic compounds, and many of them have been responsible for serious percutaneous poisoning. Particularly vicious is phenol, as well as certain substituted phenols. A report of H. M. Inspector of Factories quotes the case of a 23-year old technical assistant who was splashed by phenol in attempting to unblock a pipe-line containing the material and his trousers on both thighs became soaked. Within a few yards of a medical center, he collapsed, and despite first aid and medical treatment, did not recover. This is but one of a series of similar incidents recorded over the years, involving substantial contamination of the skin with phenol, many of which have terminated fatally.

Nitro- and halogenated phenols are probably even more hazardous and much smaller quantities can lead to systemic poisoning. Thus, low concentrations of the well-known fungicide pentachlorophenol (or its sodium salt) have been responsible for numbers of fatalities in many parts of the world. Truhaut et al. described cases of poisoning which occurred in men engaged in dipping planks of wood in a 3% solution of "penta"; two cases ended fatally.

Amines, too, are capable of being absorbed through the skin; this applies both to aromatic compounds such as aniline, dimethyl aniline, o-toluidine, diamino diphenyl methane, anisidine, benzidine and p-nitraniline as well as to aliphatic amines including butylamine, diethylene triamine, dipropylamine and triethylene tetramine.

A great variety of solvents, particularly those with good solvent power for fats, are known to penetrate the skin; among these are benzene, nitrobenzene, methanol, furfural, dioxan, dimethyl sulphoxide, dimethyl formamide, and the highly poisonous carbon di-sulphide. Not surprisingly, chlorinated solvents like carbon tetrachloride and tetrachlorethane are readily absorbed by the skin. Of even greater significance in paint formulations are the mono-ethers of ethylene glycol, such as the "cellosolves" and "oxitols". As a class, they can all penetrate intact skin; their degree of toxicity varies according to the etherifying alcohol content. Least toxic is the ethyl ether. More toxic, on the basis of published animal experiments, are the butyl, methyl and isopropyl ethers of ethylene glycol.

A number of toxic polymerizable monomers such as acrylonitrile, methyl and ethyl acrylates and acrylamide can act percutaneously but are only of any consequence if unreacted monomer should remain in the final paint. The last named monomer has been described as "dangerous". Acrylamide can penetrate the skin in toxic amounts; toxicity is cumulative. It has unusual effects on the central nervous system; the neurologic syndrome can embrace lack of muscle control and hallucinations (eye and ear).

Sometimes pesticides are incorporated into paint. Although many of these fall into two classes, halogenated hydrocarbons and organo-phosphorus compounds, the latter are not met with in paint, partly because of their instability (ease of hydrolysis) and partly because many are extremely toxic (through cholinesterase blocking). Most of the halogenated hydrocarbon pesticides are, from their chemical structure, poisonous and furthermore, most are capable of being readily absorbed by the skin. They include Aldrin, Dieldrin, Endrin and Lindane (the isomer of benzene hexachloride). Despite its good lipid solubility, however, DDT is not appreciably absorbed through the skin of man and mammals. Most organo-phosphorus pesticides are poisonous because they are cholinesterase blockers. Typical preparations are Demeton (Systox), Parathion, Malathion, EPN, Dichlorvos (DDVP), Diazinon and Phosdrin (Melvinphos).

Under (a) Ingestion, reference has been made to the toxicity rating LD_{50}. Comparable values for percutaneous poisoning are available for many compounds but are not so numerous as those for ingestion. The rabbit is more commonly used for LD_{50}(skin) determinations.

www.ingramcontent.com/pod-product-compliance
Lightning Source LLC
Chambersburg PA
CBHW080544220326
41599CB00032B/6355